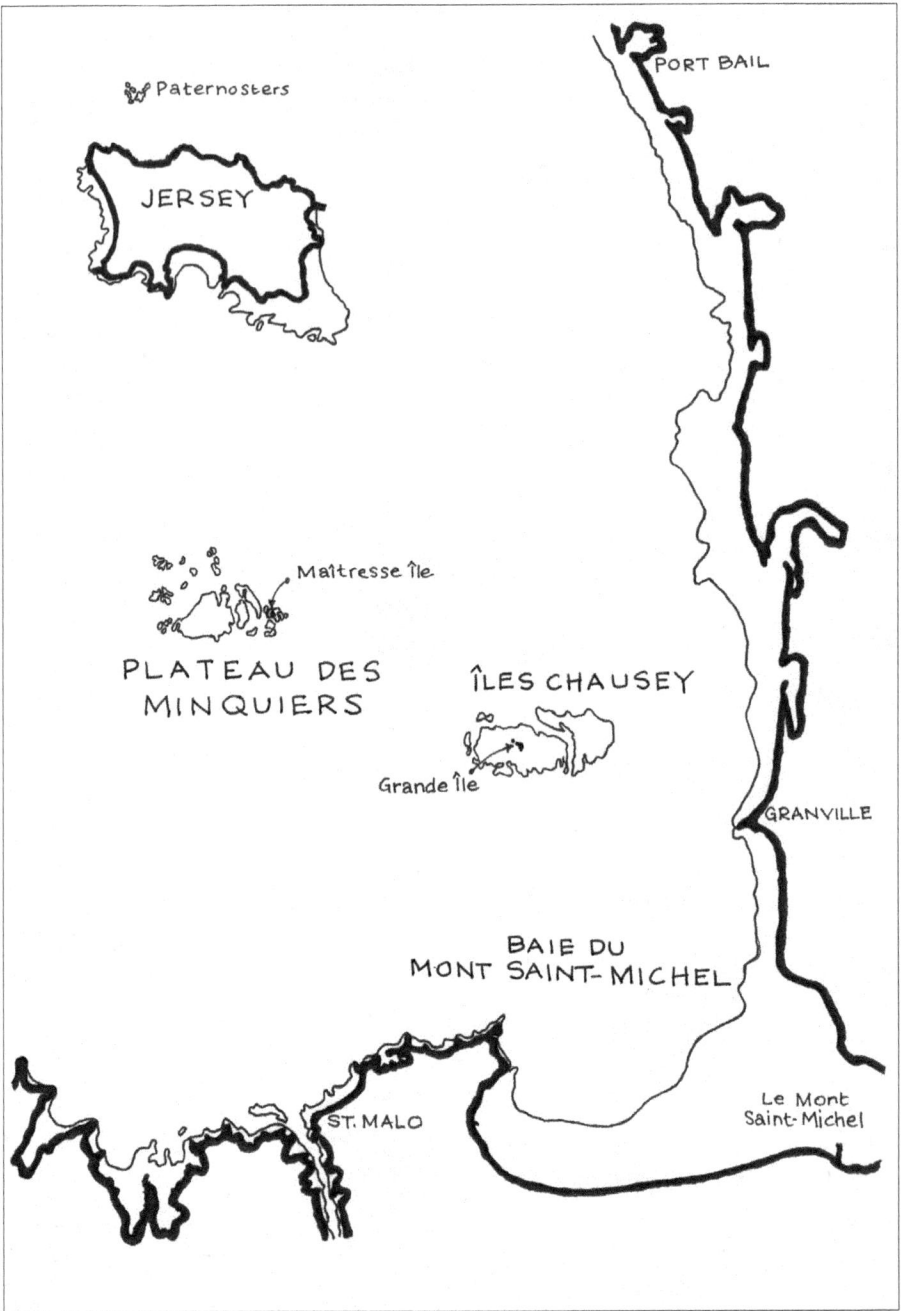

Location of Maitresse Île, Les Minquiers, in the Baie du St Michel
© AEA Design, Jersey

Les MINQUIERS

~ *Jersey's Southern Outpost* ~

Jeremy Mallinson

FOREWORD BY MIKE STENTIFORD MBE

SEAFLOWER BOOKS

First published in 2011
This revised edition published 2020 by
Seaflower Books
www.ex-librisbooks.co.uk

Origination by Seaflower Books

Printed and bound by CPI Group (UK) Ltd, Croydon, CR0 4YY

ISBN 978-1-912020-83-6

*Dedicated to the memory of my parents-in-law,
Pierre Philippe Henri Guiton (Owner of huts
'M' & 'N' on Maîtresse Île) and to Eugenie
[Ninette] Guiton (née Besnard). As well as
to my late wife, Odette.*

CONTENTS

Acknowledgements 7

Foreword by Mike Stentiford 9

By Way of Introduction 11

1 Geography and History 17

2 Archaeology 32

3 Natural History 37

4 Reefs and Wrecks 55

5 French Claims to Les Minquiers 64

6 National Flag Incidents 70

7 Guiton Family's *barraque* 78

8 Personal Landscapes 90

Bibliography 118

About the author 127

ACKNOWLEDGEMENTS

This new updated book on Les Minquiers, with special reference to La Maîtresse Île, could not have been produced without the help of many people. My thanks to my brother-in-law, Tony Guiton and to my son, Julian Mallinson, and daughter, Sophie Dixon, whose initial idea was for me to write a few notes about the gestation of the Guiton family's barraque on La Maîtresse Île. This subsequently resulted in the compilation of a much more comprehensive account of the offshore reefs of Les Minquiers.

I am most grateful to Martyn Chambers, the late Gordon Coom, Advocate Richard Falle, Juliet Gillam, Andy Hibbs, Captain Frank Lawrence, Julian Mallinson, Advocate Vincent Obbard, Paul Ostroumoff & James Painter and Martin Richardson, all hut owners on La Maîtresse Île, who allowed me to record interviews with them regarding their first involvements with Les Minquiers, and to what they consider to be the special quality of these unique offshore reefs (see Chapter 8: 'Personal Landscapes'). Also, my thanks to Mike Stentiford for writing the Foreword to this new publication about the National Trust for Jersey's 'Coastal National Park Campaign'.

My appreciation to Claude Bertram for showing me the commemorative stone on the sea wall near to his home at Fauvic, recording the names of those who aided the escapees from the German Occupation during 1944-1945, which includes the name of Peter Guiton; to George Le Couteur with regards to certain aspects of escapes to Normandy in November 1944; and for the assistance of the Connétable of Grouville, Dan Murphy.

During the research involved in compiling the book, I am grateful to the staff of the reference libraries of Société Jersiaise, Jersey's Public Library, and the archives of the Jersey Heritage Trust. My thanks to Jan Hadley for selecting a cross-section of images of Les Minquiers from the photographic archives of the *Jersey Evening Post,* and to Chris Wright and John Varcoe for granting permission to reproduce the photographs concerned. Also, my special thanks to Gareth Syvret, former photographic archivist of Société Jersiaise, for taking time to locate a varied selection of late nineteenth and early twentieth century images of Les Minquiers.

My gratitude to a number of authorities for checking and adding important data to the various Chapters. In particular to Advocate Richard Falle for allowing me to quote from his excellent text which appeared in the *History of the Parish of Grouville publication* (2001), as well as for some additional data (see Chapter 1: 'Geography and History'); to John Clarke, Chairman of the Archaeological Section of the Société Jersiaise (see Chapter 2: 'Archaeology'); to Paul Chambers, Sue Daly and Mike Stentiford (see Chapter 3: 'Natural History'); and to Tony Guiton and Julian Mallinson (see Chapter 7: 'Guiton Family's barraque').

Finally, I am grateful to Paul Chambers for his valued editorial comments and contributions to the text; to Roger Jones of Seaflower Books for adding this second edition to his list of publications; and I am forever indebted to Peter Olney for his editorial input on the 2011 publication.

I have dedicated this publication to my late parents-in-law, Peter and 'Ninnie' Guiton, and to my late wife, Odette, who introduced me to the delights of Island life, and for me to fully recognise just how privileged we have been to have had our family home on this 'Sceptred Isle' of Jersey.

FOREWORD
Jersey (Coastal) National Park

In an island needing to adjust to a rapidly changing world, Jersey's natural and maritime environment provides a desperately needed measure of stability.

While there exists a general understanding and appreciation of our surrounding natural attributes, it is nevertheless increasingly evident that sustaining such priceless environmental qualities requires every element of practical support.

Surrounded as they are by an extraordinary mobile tidal movement, Jersey's premier offshore reefs prove a fine example of such support.

Les Minquiers, Les Écréhous and Pierres de Lecq are all universally acknowledged by their internationally recognised environmental endorsements.

These protective qualifications include those of Ramsar designation and, following official political endorsement into the Island Plan 2011, as major additions to the Jersey (Coastal) National Park.

Such priority designation ensures that each of the reefs, and in particular their associated and vulnerable wildlife species, receive full and vital environmental protection.

Although each outstanding rocky outcrop is considered a gem in its own right, the remarkable storyline content relating to Les Minquiers is of dual interest and fascination.

With the newly updated 2020 publication *Les Minquiers: Jersey's Southern Outpost*, author Jeremy Mallinson is to be warmly congratulated for gathering together a wealth of contributors who enthusiastically share previously untold material relating to this remarkable and most southerly offshore reef.

From accounts of a newly established Maîtresse Île Residents Association, the bizarre raising of a Patagonian flag and of a surprising relic from the German Occupation discovered during a diving expedition; these are but a few of the reef related nuggets gracing the pleasurable pages of this new publication.

In the current light of a multitude of environmental issues, not least that of climate change, maintaining a protective instinct for all things wild and maritime has rarely proved more timely or relevant.

The voices within *Les Minquiers: Jersey's Southern Outpost* clearly endorse the collective need to support and protect the Island's most precious treasures: its reefs, its coastline, its human history and, of course, its incredible diversity of wildlife species.

Mike Stentiford MBE
April 2020

By Way of Introduction

Jersey Residency

In March 1951 I came to live in Jersey with my parents, Hal and Kay Mallinson, and my brother Miles, and since that time the family home has been in Rue du Crocquet, St Aubin. After schooling in the UK I spent 18 months working as a trainee 'vintner' at my father's wine and spirit firm in St Helier, prior to spending just over two years in the Rhodesia & Nyasaland Staff Corps in Southern Rhodesia. In May 1959 I took a summer job at Gerald Durrell's newly established Jersey Zoological Park. During my 42 years working for the conservation objectives of the Jersey Wildlife Preservation Trust (Durrell Wildlife Conservation Trust), I had the opportunity to travel widely, and to study various threatened animal species in Africa, Asia and South America.

Many aspects of Jersey have changed significantly during the last 60 years but the diversity of the Island's habitats, its many magnificent beaches, bays and cliff walks, still makes the island one of the most attractive and desirable environments in the world to live. As a devotee of the island, I fully appreciate how very fortunate I have been that my parents decided to leave Yorkshire and come to Jersey when they did, and eight years later, for my return from Africa to have coincided with Gerald Durrell establishing his Jersey Zoo. This represents very much a case of being in the right place at the right time; as a consequence, I was able to embark on a career that enabled me to spend my time and earn a living with the animal kingdom I so cherished. My career gave me the opportunity to travel extensively, and to witness the multiple problems facing the world's diminishing wildlife and wild places.

It is perhaps because of my conservation background, and having during my travels seen so much erosion of the world's biodiversity, that since my retirement I have had the opportunity to focus more on some of the conservation issues confronting this 'Sceptred Isle' of ours. Because of the States of Jersey Island Plan, which highlighted the importance of maintaining the high quality of the island's unique environment, and recognising the importance of conserving Jersey's coastal ecosystems, in the autumn of 2007 I accepted the role of Chairman of the St Aubin's Anti-Reclamation Action Group. It is my intention to continue to support all those local organisations concerned with the long-term conservation of the island's unique biodiversity and the protection of its offshore islets. In particular, The National Trust for Jersey's Save Jersey's Coastline Campaign in its mission to conserve for posterity the island's very special coastal landscapes and its associated fauna and flora.

Guiton and Minquiers Connection

In October 1963 I married Odette Guiton, sister of Yvette and Tony, whose parents were Pierre Philippe Henri Guiton and Eugenie (Ninette) Clementine (née Besnard). They were both born in Jersey in 1910 and married in October 1933 at St Thomas's Roman Catholic Church, St Helier. My father-in-law's family originated from La Rochelle, Brittany, from Huguenot stock, whereas the Besnard line came from Normandy.

Pierre Guiton, also known as Peter, was licensed as a 'Town Pilot' by the States of Jersey Harbours Department on 6 May 1933, aged 23.This was a position that he maintained until the German Occupation in 1940, when he purchased a farm at Le Boulivot, Grouville. Either on 16 or 17 June 1940 Peter Guiton participated in the evacuation of St Malo. As Le Scelleur (2000) records:

'On Sunday16 June the Bailiff of Jersey, Sir Alexander Coutanche, had received a phone call from the Governor of Jersey, Major General J.M. Hanion, relaying a request from His Majesty's Admiralty for Jersey to send all available craft to St Malo to help with the evacuation of British servicemen. The first convoy of five St Helier Yacht Club boats left at 23.00 hrs on 16 June, whereas the second group of thirteen craft left the following morning'.

Peter Guiton, in his capacity as a sea pilot, was involved in this 'Mini-Dunkirk' from St Malo, either on his future brother-in-law Denny Mourant's boat Girl Joyce, or on one of the Town Pilots boats.

The Allied Forces landed in Normandy on D-Day (6 June 1944), and by early August the Allies had broken out from the Normandy bridgehead.

A suggestion was made that if a small group from Jersey could reach France with information on the German troop concentrations and gun positions on the Island, these facts could be of great use to the Allies. As Roy Thomas (1992) recalls:

'This suggestion was received with enthusiasm, and eight individuals were chosen to carry out this mission …. Information on navigation was supplied by Peter Guiton, a States' Harbour Pilot (who was instrumental in a number of escape attempts) and John Picot from St Ouen. Such information was then 'pooled' and collated.'

Victor Huelin, Bobby Woods and Harry McFarlane made their successful escape to Normandy in a 12ft clinker-built dinghy, on a spring tide, during the night of 9 October from the Royal Bay of Grouville at Fauvic. As Roy Thomas records:

'Mr Huelin [Victor Huelin's father], being of a seafaring nature, the owner of a large yacht, was well aware of the tidal condition around the island with its treacherous currents, fast run of tide and extremely rocky coastline, arranged that they should meet the much respected States of Jersey pilot, Peter Guiton. He instructed them on the tides, the currents, and how to box a compass and he also suggested the best time of month to attempt an escape and a suitable embarkation point'.

During the night of their escape: 'Following Pilot Guiton's instructions, they rowed for a couple of hours before using the motor' (Thomas, 1992).

After the Liberation on 9 May 1945, Pierre Guiton's name was once more recorded on the States Harbours Department Registre de Pilotes until his resignation as a Town Pilot in September of that same year (States of Jersey, 1947). Possibly his last involvement with the war was in July 1945, with the boat that Messrs Kenneth Parris, Michael Price, George Le Couteur, Eric Prain and Francis Le Sueur used in November 1944 to successfully escape to Normandy. As Thomas (1992) records:

'Francis Le Sueur arranged to have Ken Parris' boat towed back to Jersey from Normandy by a fishing boat and during the journey back, the seams of the boat opened and, when approaching St Aubin's Bay, such was its waterlogged state that it had to be assisted back into harbour by the States' Pilots, Peter Guiton and Silver Le Riche.'

During 1945 the Admiralty awarded the St Helier Yacht Club a Battle Honour which gave the Club permission to fly a defaced Red Ensign. When Princess Elizabeth and the Duke of Edinburgh visited Jersey in June 1949, the Duke had especially requested a visit to the Club in order to meet those who had taken part in the St Malo evacuation, so that he could personally thank them for their action (see Le Scelleur, 2000).

On 9 May 2005, the 60th Anniversary of the Liberation of Jersey, the Connétable of Grouville, D.J. Murphy, unveiled a commemorative stone positioned on the sea wall overlooking the Royal Bay of Grouville, near to where the successful escapes to France were made. The inscription reads:

> THIS STONE WAS LAID TO COMMEMORATE THE
> HEROISMOF THOSE WHO AIDED THE ESCAPEES
> AT RISK OF THEIR LIVES
> 1944-1945
>
> BERTRAM FAMILY – Bellair   SID LE CLERCQ
> BERTRAM FAMILY – Eastlynne COLIN & BETTY MARIE
> PETER GUITON   DENNIS RYAN
> LEWIS HUELIN
>
> D.J. Murphy, Connétable 9th May 2005

It was soon after the German occupation that Peter Guiton acquired two ruined huts on La Maîtresse Île, Les Minquiers, from Lieutenant-Commander Le Breuilly of this publication. And a few years later he built a hut on La Marmotière Île, Les Écréhous.

Warwick Rodwell's excellent book, *Les Écréhous, Jersey: The History and Archaeology of a Channel Islands Archipelago*, records the following references to the Guiton hut:

'Pictured evidence dating from the late nineteenth century shows that La Marmotière boasted little more than a handful of fisherman's huts, and there were not dissimilar to the huts on La Maître Île. Hut 2 on La Marmotière was built by Pierre Philippe Henri Guiton in 1951. He sold it in 1955 for £800, without guarantee, to Alice Florence Blacker-Douglas (later, the Hon. Mrs. Westonra). On her death the hut was passed to her daughter, Cynthia Miles. The sale recorded that Pierre Guiton had obtained building consent from the 'Beautés Naturelle Comité' The structure is rendered blockwork, with a Welsh slate roof, there being a substantial concrete balcony on the south.'

A general plan of La Marmotiére, based on a survey by David Barlow and updated in 1995, is shown in Fig. 171 of Rodwell's publication; Fig. 172 includes the hut built by Pierre Guiton in 1951 (see Rodwell, 1996).

In 2007, when my brother-in-law Tony Guiton and my two children, Julian Mallinson and Sophie Dixon, embarked on the challenging task rebuilding the two ruins referred to as Huts 'M' & 'N' on Les Minquiers, I casually remarked I would research a few facts about these isolated offshore reefs for them. Also, if I did manage to find enough data about Les Minquiers that I would compile a few notes on the history and environmental quality of the La Maîtresse Île. However, the more I started to delve into what already had been written about Les Minquiers, the more fascinated and enthusiastic I became about them. So after a considerable amount of research in the libraries of the Société Jersiaise, Jersey's Public Library, the Jersey Archives, and talking to a number of people with considerable knowledge about the reefs, and in particular La Maîtresse Île, what started off as a collection of a few notes of information gradually evolved into my first 2011 publication about the Minquiers.

However, since writing the text some ten-year's ago, a good number of interesting events have taken place which include: the establishment of a 'La Maîtresse Île Residents Association'; new conservation legislation; a note on a further wreck; the third hoisting of a Patagonian flag; hut ownerships; additional natural history observations; new photographic images; some fifty additions to the Bibliography; and in 2016, an excellent publication of a comprehensive scientific appraisal of *The Natural History of Les Minquiers*, by Paul Chambers, Francis Binney

& Gareth Jeffreys.

In summary, these isolated offshore reefs of the Minquiers group of islets could quite well be described as representing a Marine Venus that deserves future total protection, so that a unique and important ecosystem may be preserved in its entirety for posterity.

Chapter 1

Geography and History

Richard Falle's contribution to the Parish of Grouville's excellent publication: *Grouville, Jersey - The History of a Country Parish*, well covers both the geography and history of the Minquiers reef. I have therefore selected from his chapter on Les Minquiers the most pertinent to this publication, and unless otherwise cross-referenced, the text follows Falle (2001) and is apostrophized ('… ') accordingly.

During the Ice Age there was no sea between the French coast, Jersey and its offshore reefs. About 8000 years ago sea levels rose and Jersey became an island, and Les Minquiers and Les Écréhous became reefs. Geographically, the plateau of Les Minquiers lies 12 miles due south of Jersey and measures, from Les Brisants du Nord Ouest to Les Grelets in the east, an extreme length of 18 miles; its breadth from north to south, reaches a maximum of 14 miles. The reef marks the most southerly point of the British Isles. Known by Channel Islanders as "Minkies", which as Jefferson (1985) records: 'is derived from an old French word, minkier, which means a fish wholesaler, and relates to the abundance of fish found in these waters. It is part of the Bailiwick of the island of Jersey and has long historical connections with the Parish of Grouville, the Vingtaine of La Rocque and Fief of Noirmont'. As Richard Falle records:

'The Minquiers lie deep in the great bay defined on the east by the Cotentin peninsula and on the south by Brittany. It is known variously as La Bai de Granville or Baie du St Michel. Twelve miles south of Jersey, the reef is a little more distant from the adjacent French coast by a mere eight miles north of Chausey. Much of the reef is hidden at high tide but with the ebb and flow of the huge tides the appearance of the reef is constantly and dramatically changing. At low water springs, a large plateau is revealed. On the high tides the plateau all but disappears leaving only a small scattering of rocky heads barely showing above the surface of the water, at the western end a few small stacks called Les Maisons, face the Atlantic. The turbulence of the waters in these western approaches is marked on the charts "overfalls", a term pregnant with meaning and calculated to deter most prudent mariners. At the east end of the reef lies Maîtresse Île, at high tide, barely a hundred yards long and half as wide. The little islet with its surrounding rocks provide a relatively safe haven for yachts and fishing boats, and a site for about a dozen or so huts or barraques. These are mostly built with the honey coloured gneiss of the reef whose subtle harmony is but slightly disturbed by the two or three buildings of brightly coloured granites imported from Jersey and France.'

'When the tide retreats it uncovers a great expanse of sharp rocks, lagoons and rivers and gloriously curved banks of sand, gravel and shells. It is horizontal world; nothing stands high above sea level. On a clear day from high ground on Jersey's coast, the Minquiers appear as a long low string of rocks stretching along the southern horizon. Sometimes when the atmospheric conditions allow, the rocks seem to rise up and miraculously float on or just above the sea.'

In the history of La Maîtresse Île:

'Recent activity on the island has led to the erosion of the soil cover. The evidence of fires in the presence of 'pot boilers' and seal bones suggests that at the end of the last ice age when sea levels were lower, La Maîtresse Île was the habitation of Neolithic seal hunters. In historic times there is however, little in writing concerning the Minquiers before the 17th century. Prior to that, inferences have to be drawn from the records of Chausey which lies only eight miles to the south of the Minquiers. It is an archipelago of some fifty islets and represents the only French part of the Channel Islands. Its records dating from medieval times have been helpful in filling the gaps in the history of the Minquiers.'

These archives recorded:

'The Abbé du Mont St Michel en Peril de la Mer was once lord of Chausey.

His title dated from a grant made by the Duke Richard 11 of Normandy in 1022. This grant included land in Jersey, parts of the Contentin and the "insula que dicitur Calsoi" (the Island called Calsoi). Perhaps the Minquiers were deemed included with Chausey. It seemed certain that the Abbé in any event, regarded the Minquiers as his. Such a claim to the Minquiers and the Écréhous was subject of France's claim to their sovereignty at the International Court of Justice in 1953.' (see Chapter 5 of this publication: French Claims to Les Minquiers.)

In recent correspondence Richard Falle emphasised:

'that Chausey and therefore the Minquiers to the north were subject to the English Crown at least until the 16th Century. There being at least three proofs of this assertion (a) The Assize held in Jersey in 1309 summoned the Abbot of Mont St Michel to answer on a number of matters. It is recorded that Chausey was then in his possession. This would hardly have been material if the Reefs were not considered to be part of the English King's territory, (b) Some fifteen years later the same Abbot was before the French King's Court in a dispute when it was accepted that the French King had no jurisdiction over Chausey, and (c) The Anglo-Norman Islands were held by the English King for centuries after 1204 but remained in the diocese of Coutances. This anomaly was finally resolved in 1500 when, by the Bull of Pope Alexander VI, the Islands expressly including Chausey, were transferred to the Diocese of Winchester. Although not indicated, it must be clear that the Minquiers would have been included in this document of international significance' (Falle, pers. comm.).

When King John in 1204 lost continental Normandy to the French he was determined to hold Les Îles which, sitting astride an important trade route, were vital to communication between England and John's vast possessions in Gascony.

'Throughout the long period following the loss of continental Normandy, until the 16th Century, the Anglo- Norman Islands remained subject to the jurisdiction of the Bishop of Coutances within whose diocese they lay. Large parts of the Islands remained the property of the great French monastic houses. Among these the Abbé of Mont St Michel possessed Chausey, the Minquiers, and in Jersey the Fief of Noirmont, the Priory and his fief in St Clément and, in Grouville, the Fiefs of Dameraine and Aumosne. The chief representative of the Abbé in Jersey was the Prieur de St Clément. The Abbot's lands in the Islands were confiscated by Henry V in 1413 under a process known as the "Seizure of the Alien Priories". Thereafter, the Minquiers were in the hands of the English Crown.'

'The feudal relationship between the Abbé and his various tenants would have given the men from Jersey the run of the fishing on the Minquiers reef. Most of the fishermen living in the low stone cottages along the shoreline at La Rocque looked in the Middle Ages and after upon the Minquiers as their exclusive fishing grounds. In 1643, George de Carteret, captain of one of Her Majesty's ships, rescued some English sailors from Moorish pirates on the North African coast, and was rewarded by Charles I with a grant of Letters Patent of the Jersey fiefs of Norman, Melesches and Grainville. This grant and the various Royal confirmations of title given to successive siegneurs of Noirmont down to the present day conveyed to them all the ancient rights which had been enjoyed in centuries before by the Abbé of Mont St Michel. In the surviving court rolls of the Fief of Noirmont covering the late 16th and early 17th Centuries there are references to the Minquiers. These deal with seigneurial matters relating to wrecks and the exercise of jurisdiction by the manorial court, and the Jersey men recorded in these cases as being on the Minquiers reef were, it seems, men of Noirmont.'

It is interesting to note that Jersey law used to be that all debris found upon the shore or floating on the sea is the property of the Crown and that, by tradition a third of the value was given to the finder (although the Crown is not obliged to do so), a third to the Seigneur of the Fief where it was found, and a third to the Crown. However, in 1966 the Seignorial rights (Abolition) (Jersey) Law cancelled the rights of Seigneurs to jetsam and flotsam so in effect the Crown now claims two-thirds of the value and the finder is still allowed one-third. C.S. Le Gros records in Traite du Droit Coutumier L'Ile de Jersey that: 'whatever comes so near that a man on his horse can attach it with his lance, must be declared to the Seigneur, who was entitled to keep it for a year and a day if the owner could not be found, after which the proceeds were divided between the Seigneur and the finder. Heavy penalties were incurred if the find was not reported and cases of persons who had not done so being thrown into prison in these times was common' (after Le Gros, 1929/1936; JEP 1969).

'The first fishermen from Jersey may well have gone to the Minquiers in pursuit of the conger. Lobsters were not mentioned until much later. In contrast, conger and to a lesser extent, mackerel, were in the Middle Ages, caught in huge quantities in the islands to be dried, salted and sold in an extensive trade as far south as Gascony. There is, however, nothing in the records to confirm such speculation in relation to the Minquiers save perhaps the reference in an early 17th century case recording a dispute before the Royal Court over the tithes due by the men of Noirmont fishing on the offshore reefs to the Rector of the Parish of St Brelade. It seems that the first fishermen at the Minquiers used to sleep on their boats, and later stayed in the small stone barraques that had roofs thatched for shelter with straw and dried vraic.'

As Richard Falle points out:

'the rock of the reefs, being easily worked and accessible from the sea, represented a valuable asset as a quarry for building stone. Large parts of the magnificent construction on Mont St Michel (La Merveille) and the town of Granville on the adjacent French coast are built of this material. So too are parts of 15th century buildings in Jersey, including the gargoyles in Grouville Church. The characteristic blue from Chausey is also to be seen in many 17th and 18th century Jersey houses. If the Abbot of Mont St Michel quarried Chausey he probably also quarried the Minquiers. In contrast to the rather cold blue of the Chausey granites, the gneiss of the Minquiers has a warmth and softness in colour not unlike the paler browns of Mont Mado and much less brash than the pinks and reds from some of the other granite quarries in Jersey. It is speculated: "Could it be that some of the browner stone in old buildings described as Chausey, in fact comes from the Minquiers?" Certainly, physical inspection of the Minquiers reef shows evidence of the quarryman's activity everywhere.'

'Unlike the Écréhous, where some of the structures date back to the

13th century and perhaps earlier, the oldest huts on La Maîtresse Île are of more recent origin. Only one hut is shown on the sketch which accompanies Captain Martin White's [Royal Navy] cartographical survey of the reef in 1812. He wrote: "There is a hut built on the Island for the occasional protection of the fishermen and vrachers (gatherers of seaweed) who frequent the place for the purpose of obtaining the conger, ormer (oreille de mer) and lobsters which abound in great profusion". However, it seems more probable that the stone huts on La Maîtresse Île were built by quarrymen cutting rock for the construction of Fort Regent [Jersey] at the beginning of the nineteenth century' (after Falle, pers. comm.).

La Maîtresse île, boulder with quarrymen's initials
Photograph Ref: SJPA/013099, 1896 © Société Jersiaise

Godfray (1929) records: 'how quarrymen have cut their initials in many places on the Maîtresse Île, and that a record had been made of some of these. The oldest date found is 1792 and to assign the oldest of the stone huts to approximately the same date would be in conformity with such oral tradition as it was possible to obtain from the present fishermen. In the latter part of the 18th and early 19th centuries the Crown looking for readily accessible building stone, for the fortification of the Town Hill (Mont de la Ville, Jersey), sent a team of quarrymen to La Maîtresse Île. The rock from that island, and probably from other parts of the reef was cut and easily floated over the sea north to St Helier by barge, for the building of Fort Regent'. The foundation stone for which was laid on 7th November 1806 (see Davies, 1971).

'The extensive quarrying that took place was not without controversy. The granite was mainly taken from the northern end of the island, with drill and blasting powder, and the quarrymen had already removed much stone before being stopped by the irate fishermen. First by the simple and effective method of having all their tools taken away and dropped into deep water, and secondly by the more constitutional procedure of the fishermen laying a petition before the responsible authorities. In consequence 'La Pointe' still remains as a bulwark against the sea and a wall has been built to prevent further encroachment from the south-west. If the quarrying had been permitted to proceed much further, the whole island would now be barren.'

It had been during this period that some small granite dwellings had sprung up to accommodate the workforce, but the official protestation to the States of Jersey in 1807 brought about the eventual demise of the quarrying industry on the reef. When the quarrymen abandoned their huts they were almost certainly then taken over without any formality by the fishermen and the reefs reverted to their more familiar role as fishing grounds, with their harvesting traditionally the prerogative of Grouville fishermen. This long association of the Parish of Grouville with the Minquiers evolved into a formal administrative responsibility and to this day parochial matters pertaining to the locality, rate-assessment and policing, fall within the jurisdiction of the Parish and its elected officers. As the nineteenth century progressed and the importance and volume of fishing increased, re-development of La Maîtresse Île took place and the abandoned cottages were renovated or rebuilt. By 1888 the Jersey Piers and Harbours Committee recorded 18 inhabitable houses regularly used by visiting fishermen. A slipway was constructed by the States in 1907 and lengthened in 1933 (after Butlin-Baker, 1981; Falle, pers. comm.).

During the Second World War the German occupying forces stationed an anti-aircraft battery on La Maîtresse Île, as well as the island

Thatched cottages on La Maîtresse Île and beacons
Photograph Ref: SJPA/013112, 1885-1895 © Société Jersiaise

Cottages on La Maîtresse Île
Photograph Ref: SJPA/013101, 1893 © Société Jersiaise

Shrimping at Les Minquiers
Photograph Ref: SJPA/05340, 1895-1905 © Société Jersiaise   Photographer Albert Smith

Cottages and flagpost on La Maîtresse Île
Photograph Ref: SJPA/009131, 1896 © Société Jersiaise

La Maîtresse Île, cottages and rocks
Photograph Ref: SJPA/011451, 1915-1925 © Société Jersiaise

La Maîtresse Île, fishermen with nets and lobster pot
Photograph Ref: SJPA/013115, 1913 © Société Jersiaise

La Maîtresse Île cottages, people and sea wall
Photograph Ref: SJPA/013095, 1915-1925 © Société Jersiaise

La Maîtresse Île, view between cottages, men in distance
Photograph Ref: SJPA/013088,
Photographer Emile F. Guiton, 1928 ©
Société Jersiaise

View of La Maîtresse Île with cottages and the States' tug,
the Duke of Normandy
Photograph Ref: SJPA/013082. Photographer Emile F. Guiton, 1928
© Société Jersiaise

View of La Maîtresse Île cottages (east-side) showing steps and lobster pots
Photograph Ref: SJPA/013106, 1939 © Société Jersiaise

functioning as an observation post. As Butlin-Baker (1981) records: 'It was the most isolated and probably the most unpopular posting in the Channel Islands, and during their dutiful vigil for an invasion which never materialised, did little to change permanently the tiny island, although they did strip all but two of the buildings for firewood and succeeded in damaging a flagstaff erected in 1890.'

During the war years the German authorities prohibited Jersey men from fishing on the Minquiers. However, as soon as the Allies had liberated the Continent, French fishermen descended on the reef in large numbers while Jersey remained under the 'jackboot' for almost a further year. This accident of history was to have momentous effects for the Jersey fishing community. For after the German Occupation of the Channel Islands, Messrs Le Clerq and Gallichan sought to return to their ancestral fishing grounds, but they found the French in possession. They were later to say that they were, in effect, elbowed out. Faced with superior numbers of French in larger boats, determined to hold on to their rich new fishing grounds, the Jersey fishermen were intimidated, and it was a great sadness for them and their families when they had to take-up other occupations (after Falle, 2001; Lemprière 1970).

The non-flushing toilet on La Maîtresse Île, the most southerly building in the British Isles, was constructed by the States of Jersey in the 1930s, although in Captain 'Pull-through' Bolitho's time (the previous owner of Vincent Obbard's nearby hut) the latrine was frequently referred to as 'Pull-through's loo' and also used by him as a shipping mark (Obbard, pers comm.). However, it is interesting to note the furore caused 60 years later when Jersey Customs and Excise (Impôts) Department organised a Territorial Army unit to build, along-side the Customs hut on the Écréhous, a flushing lavatory, water tank and outlet. A press report in the *Sunday Telegraph* titled: 'First flushing toilet panned by islanders' recorded: 'Development on such an extravagant scale has incurred the wrath of Jersey families with holiday retreats on the Ecréhous, including retired Brigadier Raoul Lemprière-Robin … The Brigadier and his friends want the water tank moved -"The toilet is now the grandest structure on the Écréhous," he said. "To put it bluntly, it has been a bugger's muddle all along." The report went on to record the president of Jersey's Island Development Committee, John Le Sueur, saying that the development may be reconsidered 'He knows the feelings of the Écréhous summer residents are not to be taken lightly. "They are incensed." He said. "They guard going to the Écréhous jealously. I'm from a family that goes back to 1400 [in Jersey] but I wouldn't dare go on the rock" (Mourant, Andrew, 1992).

The most southerly toilet in the British Isles
Photographer Tony Pike, 19 September, 1996 © Jersey Evening Post

A letter written by Vivian Richardson, dated 22 January 1977, provided information on the then ownership of what he referred to as 'cabins', on La Maîtresse Île. Attached to the letter was a rough sketch recording previous and current owners of 21 different sites. Apart from the mainly repaired Vincent Obbard hut on the south; the States Impôt building (accommodation, storeroom and yard); the newer States Harbours & Airport building; and William Coom's hut on the north, generally known as L'Hôpital; the condition of the majority of the other structures on La Maîtresse were recorded as being in a 'v. ruinous' state (M. Richardson, in litt.).

Certain buildings were given to the States in the 1st January 1922 contract passed before the Royal Court in which it states that the properties were given to the then Public Works Committee. The States own the Impôt Hut and Yard; the States House (formerly Harbours &

Airports), which was designed in 1938 and finished in 1945; the Public Toilet; the Helicopter Pad constructed in 1969 on the site of a hut once owned by Sir Robert Masurier, a former Bailiff of Jersey; the Flag Pole, Slipway; and Beacons/Masts. Harbours and Airport staff maintain the buoys and beacons and keep the slipway in repair (after Sweeney, in litt. 2009). 'By tradition, whenever a fishing party descends on the island, the first ritual is the raising of the Union Flag: as assertion, perhaps old fashioned, of national identity, on this compromised but beautiful frontier' (after Falle, 2001).

Chapter 2

Archaeology

A.D.B.Godfray's archaeological researches on La Maîtresse Île are comprehensively recorded in his Société Jersiaise publications, so the following text, up to references to further archaeological visits in 1980 and 2006, are taken from his observations (Godfray, 1929, 1931):

'Of this great plateau, only nine heads of rock are not submerged at high water; of these, eight are entirely barren of vegetation, but the remaining one, the largest, still retains a variety of plant life remarkable for so small an area. It is known as La Maîtresse Île and is shaped roughly as a Y, of which the stem points to the north. It is approximately 100 yards long, by 50 yards wide; the extremities of the stem of the two branches are barren except where thrift and a fine-leafed couch grass manage to retain a shallow turf, but the fork of the Y, which is also the lowest part of the island, holds in places as much as four feet of soil and it is here, also, that the greatest number of fisherman's huts have been built.'

As previously mentioned, at the end of the 18th and beginning of the 19th centuries, an enormous quantity of stone has been quarried from La Maîtresse Île and its immediate neighbourhood. both by blasting and by the older method of wedging off blocks along the natural splits of the rock, to which the simple form of attack the Minquiers granite readily yields. These blocks lie in hundreds all round the island and many of them bear the V shaped cuts which are the preliminary to further splitting. The result of this quarrying has been to admit the sea, particularly from the south west (i.e. up the fork of the Y) during stormy weather, with most deplorable results, for a great part of the richest soil, as the archaeologist measures wealth, has been washed away. Indeed, if the quarrying had been permitted to proceed much further in the early 19th century, the whole island would now be barren (see Chapter 1: Geography and History, in this publication).

During Godfray's July 1928 excavations:

'Two trenches about eight ft long and three ft wide were opened in two areas: one to the north of the slipway and east of the building known as 'L'Hôpital', and the other immediately to the north west of the Impôt house, where investigations were carried out. The information obtained from these was identical and in neglecting the overlay of rubbish accumulated during comparatively recent times, two archaeological horizons, entirely distinct and separated from each other by a wide interval of time, were found. The upper horizon consisting of from three to six inches of black sand yielding polished pottery and bones in large numbers; the lower horizon, separated from the upper by six inches of white sand, occupied only the upper few inches of closely-packed black pasty soil which lay nearest to the rock, and which varied in thickness, but nowhere appeared to exceed 18 inches. This lower horizon yielded flint knives and scrapers of good workmanship, 'fabricators' or long, sea-worn pebbles of convenient size, applied to many uses, and bones in considerable quantities, the majority being those of Seals, which predominate in the upper horizon.'

'Polishing as a means of hardening the surface and body of a pot is a very ancient practice and many evidences of it have been found which shows that it was common in Jersey in early Bronze Age and late Neolithic times. The Minquiers pottery had been fired at approximately the same temperature as the late Neolithic and so it would appear to lie, in point of time and development, somewhere between the Jersey Bronze and the early Norman pot found in St Helier's churchyard and its age was estimated to be placed at 2000 years. The available evidence seemed to show that the denudation of the eastern end of the plateau of the Minquiers occurred at two distinct periods; the first before the habitation horizon of 2000 years ago, while the second brought that horizon to an abrupt close.'

'The first is most clearly shown by the clean, white and sterile sand lying between the lower and upper horizons. This sand is not stratified and gives no indication of having borne any vegetation until the period of the upper horizon is reached; its appearance suggests that it all came in one short period of time as the result of some great storm demolishing a sea barrier of sand dunes and it seems to have remained loose until the vegetation of the upper horizon gradually prevailed and bound it. The shells of the Wood Snail (*Helix nemoralis*) found in the upper horizon showed by their size that the vegetation must have been far more luxurious than it is at the present day.' Godfray postulated that: 'a large area of vegetation-bearing soil must have been left in the neighbourhood of the Maîtresse Île after the first denudation.'

'The second, which eventually reduced La Maîtresse Île to its present dimensions appears to have been more gradual, for the sand shows traces of having borne vegetation throughout its period of deposition. Yet the beginning of this final period was probably catastrophic also, for there appeared to be no reason other than suddenly changed conditions to account for the abrupt cessation of the visits of the Seals (*Halichocrus grypus*) which must previously have frequented the Minquiers in vast numbers. In the natural course of events Seals die at sea and not in the rookeries, and so the great number of their bones found there must be attributed to their having been exploited by man as food. Consequently, when the seals no longer came to the Minquiers, the men who hunted them ceased to come too.'

Godfray concluded his 1929 Archaeological Report on the July 1928 researches by commenting: 'What a rich field for investigation lies waiting the archaeologist at the Maîtresse Île, and that there is also work for the geologist [see Mourant, Arthur, 1933 & 1973], the biologist [see Chambers, 2008; Daly, 2004; Le Sueur, 1976] and the historian [see Falle, 2001], and suggested that a comprehensive report on the whole plateau should be undertaken'. In a later Archaeological Report, he went on to relate that another dig had been carried out at La Maîtresse Île in July 1929. A large number of Wood Snails were found in the clean sand above and below the Upper Horizon and large quantities of coarse pottery shards were recovered. Also, a larger number of bones, practically all belonging to the Grey Seal were brought back for expert examination. The general result of the visit confirmed the previous report, which appeared in the 1929 Bulletin (Godfray, 1931).

In 1980, an expedition was made to Maîtresse Île to examine the condition of one of the Bronze and Iron Age horizons. Finlaison (1980) records:

'that little more than a pocket handkerchief-sized piece of land now remains and the help of the conservation officer has been required to solve the problem of containing the land and the last remaining pockets of stratified evidence. Loss of soil has been greatest at the northeast end of the island between flagstaff hill and the slipway, where the cliff receives the full brunt of the waves in bad weather. From here, two lenses of Bronze Age debris, principally pottery, including the rim of a cordoned urn (identified to fig. 29,d, page 115, J. Hawkes, Arch. of the CI), Limpet shells and Seal bones were found crumbling on the beach. The sea also comes right up the South Gully and on occasions breaks over the entire island. Surface finds of flint and pottery were found everywhere in the land between the long wall and the huts. No Iron Age pottery was seen and in fact the La Tené level here may be almost completely lost as the dense black horizon of the Bronze occupation is now only about five to ten cms beneath the surface.'

In the summer of 2006 a small group from the Archaeology Section of the Société Jersiaise, with its chairman, John Clarke, made two brief visits to Les Minquiers where two late 18th century fishermen's huts ['M' & 'N'] were being rebuilt. The first visit recorded:

'The form of one hut and a trench of 60 cm x 30 cm which indicated the Iron Age, and Bronze Age surfaces still exist underneath the huts. These finds were consistent with the reports of Godfray in 1928. A second visit was made, and across the adjacent hut a 1 m x 4.5 m trench was dug down onto the Bronze Age peat soil layer, where a considerable number of remains were found ... This brief visit has shown that we need to study the archaeology of the isolated and remote parts of these islands whenever the opportunity arises'
(Clarke, 2006; Clarke, 2009). Also see Chapter 7: Guiton Family's barrraque, in this publication).

Foundations of the Guiton Family's barraque
Photograph May 2006 © Julian Mallinson

Chapter 3

Natural History

As previously recorded (see Chapter Two), archaeological excavations on Les Minquiers have shown that 2,000 years ago, and even earlier, Grey Seals were abundant on La Maîtresse Île and men went there to hunt them but it is thought that later the rise of sea level destroyed this ideal seal habitat. For centuries French and Jersey fishermen shared the harvest of the Minquiers, their harsh lives being immortalised in Victor Hugo's famous novel *Toilers of the Sea* (see Carnegie, 2006). During the 18th and 19th centuries a small community of fishermen from La Rocque used Maîtresse Île as a base between Monday and Friday, rowing or sailing their boats out to the reef and returning to Jersey in time to catch the market with their great catches of lobsters, Ormers (*Haliotis tuberculata*), Congers (*Conger conger*) and other fish (see Lemprière-Robin & Falle, 1986). As Paul Chambers (2008) records:

> 'The Channel Islands are subject to some of the largest tidal movements in the world with ranges that vary from 12.2 m in Jersey to 6.7m in Alderney. This generates strong tidal streams which, during spring tides, may flow at up to five knots through 'the Race' between Alderney and the Cherbourg peninsula, but which regularly exceed three knots around the other islands. Jersey, Les Minquiers and Les Écréhous are situated on a shallow shelf (less than 30 metres deep) that is associated with the Gulf of St Malo, while Guernsey, Sark and Alderney are surrounded by deeper water of 40 m or more'. The strong currents, larger tides, shallow waters and an exposure to westerly gales often causes the islands to have high turbidity; these same factors may also cause the seas to become nutrient-rich and subject to seasonal planktonic blooms; a phenomenon which provides food for many organisms, including molluscs. The

total number of molluscs species recorded for all the Channel Islands is 440, whereas the number recorded for Les Minquiers is 89, and for Les Écréhous 34.'

The rocky, fragmented nature of this landscape means that it is subject daily to fast-flowing currents that rip through the reef at speeds of up to three knots. At high water the lack of land means that it is very much exposed to action of wind and waves, including the worst effects of Atlantic storms. This makes Les Minquiers a harsh water-swept terrain dominated by bare rock and coarse sands and gravels. It is in these rather restricted habitats that the marine life must make itself at home, but this is no easy task. Considering its large area, Les Minquiers contains a relatively restricted number of habitats in which plants and animals can survive. The most important of these are the igneous and metamorphic rocks which dominate the reefs both above and below the water (after P. Chambers, pers. comm.). Chambers went on to state:

'The bare surface of these rocks provides some purchase for plants and animals, especially near the low water mark where large seaweeds such as *Laminaria*, *Sargassum* and *Saccorhiza* can form themselves into marginal marine forests that, in turn, provide shelter and food for many other species. The erosional power of the sea has fashioned the rocks so that many contain overhangs, gulleys and deep, narrow crevices. These provide shelter from both the currents and direct sunlight and may contain a wealth of plants and animals, including such fishermen's favourites as the Lobster (*Homarus gammarus*) and the Ormer. In times gone past these areas also sheltered vast numbers of the common Octopus which would breed in the shallow waters of the reef and then spread north to Jersey and, in plague years, to the south coast of England.'

'The sediment within Les Minquiers is also reflective of its topography and oceanography. Much of it is coarse sand or gravel and there is little silt, clay or other fine-grained sediment in the reef. The centre of the reef is home to a series of tall, golden sandbanks which are built from accumulations of gravel, sand and shell debris that is washed into the reef and deposited at high tide. These sandbanks are coarse, mobile and saturated by seawater; few species can tolerate such conditions but on large tides the lower reaches of the banks do reveal a moderate number of large bivalve molluscs including Razorfish (*Ensis spp.*), surfclams (*Spisula spp.* and *Donax spp.*) and the once much sought after Five Shilling Shell (*Mactra glauca*). Rarer areas of fine sand and silt, such as the anchorage at Maîtresse Île, show a greater diversity of sediment-dwelling species, including many smaller molluscs, worms and burrowing echinoderms.'

'In the waters that lie above the seabed lives a variety of fish and other swimming organisms. The reef is famous among sea anglers for its sizeable Sea Bass (*Dicentrarchus labrax*) but a variety of other edible fish live there also, including various species of wrasse, mackerel, mullet and others. The nooks and crannies of the reef and their associated seaweed forests provide shelter for fish and their prey while the sandbanks are stocked with large number of sand-eels. In addition to smaller species, Les Minquiers is anecdotally known as a place where large sharks may be found. The biggest of these, the Porbeagle Shark (*Lamna nasus*), lives in the deeper waters around the edge of the reef but in times gone by some large specimens have been caught including the world record specimen of 430 lbs which was taken by Des Bougourd in 1969. The shallower water is home to Tope (*Galeorhinus galeus*) and various species of dogfish' (P. Chambers, pers. comm.).

A brief report from 1926 on 'La Faune de l'Archipel des Minquiers', published in the *Bulletin du Museum d'Histoire Naturelle*, Paris, found a total of eighty-one species of molluscs (Fischer, P. & E., 1926).

Up to the 1970s no annual surveys on bird populations, or studies of bird migration, had been undertaken on either Les Écréhous or Les Minquiers. Although the value of carrying out such studies had been highlighted, particularly with migratory rarities, such as the three Nightingales (*Luscinia megarhynchos*) seen on Les Minquiers on 17 August 1958 (Le Sueur, 1976 - see plate 17).

Birds have always been prolific on Les Minquiers and Les Écréhous. In 1937 nests of both Cormorants (*Phalacrocorax carbo*) and Shags (*P. aristotelis*) were photographed by Roderick Dobson on the Minquiers reef, and Dobson recorded that Maître Île within Les Écréhous represented a frequent breeding place for gulls and terns, Shags and many others; and that in all probability Les Minquiers had similar suitable nesting sites for these species (see Dobson, 1952). In 1952, members of the Société Jersiaise Bird Section, under the guidance of Frances Le Sueur, erected a net trap on the Minquiers to study the migration of birds, work that was recognised by the British Trust for Ornithology (JEP Temps passé, 2004 – see Plate 18). In 1959, an addled egg of a Storm Petrel (*Hydrobates polagicus*) was found on the Isle (Le Sueur, 1976).

with Frances Le Sueur in Spring 1972 (the two Guiton family's ruins in centre of picture) © *Jersey Evening Post*

A report on a visit to Les Minquiers on 26/27 July 1973 by the Joint Nature Conservation Advisory Panel and the Société Jersiaise recorded: 101 Shag nests and other breeding birds including Herring Gull (*Larus argentatus*), Great and Lesser Black-backed Gull (*L. marinus & L. fuscus*), Oystercatcher (*Haematopus ostralegus*), Storm Petrel and Starling (*Sturnus vulgaris*). Non-breeding and migratory bird species included: Sandwich Tern (*Sterna sandvicensis*), Gannet (*Sula bassana*), Turnstone (*Arenaria interpres*), Redshank (*Tringa totanus*), Blackbird (*Turdus merula*), Leaf Warbler (*Phylloscopus spp.*), Rock Pipit (*Anthus spinoletta*), and Common Tern (*Sterna hirundo*) and Sandwich Tern (*S. sandivensis*) (Le Sueur et al. 1973). In 1976, Shag's nest was observed inside a derelict hut on Les

Minquiers, and Cormorant nests were evident on the high flat-topped rock at the south-west end of La Maîtresse Île, and on Les Maison at Les Minquiers reef (see Le Sueur, 1976).

In the latter part of the 1970s Juliet Gillam recalls that early in the year the whole offshore reef reeked of guano and there were birds everywhere. 'I remember Bob Le Masurier's derelict hut (where the Helicopter pad now is) being packed full of young Shags. There were Shags everywhere, a lot of Herring Gulls, Oystercatchers, but very few Great Black-backed Gulls. Unfortunately, now with so much human activity and the lack of ruins, there are very few places for them to nest' (J. Gillam, pers. comm.).

Roderick Dobson (1952) recorded that the Red Kite (*Milvus m. milvus*) was still nesting in Jersey up to about 1815, but since that time became a rare vagrant. However, on 6 October 2016 it was reported in the JEP that a red kite had been observed as a rare visitor to the Minquiers, but this was subsequently dismissed by the ornithological section of Société Jersiaise. Although in November 2016 possible sightings of a red kite in the Channel Islands were reported more widely the first since the 19th century. Prior to these, there had only been eleven records of red kites n Jersey, and these recent sightings are likely to have resulted from the highly successful reintroduction project in the UK (Young, 2016 & Young, pers comm.).

During the period 30 June to 5 July 1992 a team from the Marine Laboratory of the University of Portsmouth carried out a survey on the marine biology of La Maîtresse Île. Before this, very few publications refer to the marine biology of the islet, and this remains the only study of Les Minquiers to provide any detailed ecological data. However, as in all such initial investigations of an area the survey team recognised that its priority was to initiate documentation of the region and to determine what biota were present. Appendix 1 of the Survey Report provided a listing of 150 identified species of marine plants; and Appendix 2 recorded a listing of marine animals which included 91 animals identified at least to genus, and ten more recorded to group. In total, the Portsmouth team identified 225 species from all the main animal and plant groups, although the Survey Report emphasised that the work undertaken was confined to La Maîtresse Île and therefore represented only the beginning of what needs to be a much more extensive survey (after Culley et al. 1993).

As a follow-up to the Portsmouth University survey of La Maîtresse, recent studies by Paul Chambers have led to 91 species being recorded from the reef:

'These are mostly molluscs but in 2009 a large tongue-eating parasitic isopod (*Ceratothoa steindachneri* - a woodlouse-like creature) was recovered from the mouth of a Lesser Weever Fish (*Echiichthys vipera*). It was only the second British record for this species and its unusual lifestyle attracted media coverage worldwide. A database of Channel Island marine species [maintained by P. Chambers] holds 553 records from Les Minquiers, taken from a variety of published and first-hand sources. This equates to a total of 339 marine species recorded from Les Minquiers, which compares with some 2,700 species known from the Channel Island region as a whole. Amongst this total are some relative rarities, including the Brittlestar (*Ophiopsila aranea*) whose only British record is from the reef. It is expected that further survey work will be undertaken in the coming years both by P. Chambers and by the Marine Biology Section of La Société Jersiaise (P. Chambers, pers. comm.).

A satellite tagging system developed in Scotland has enabled French marine biologists from Oceanopolis, an educational and research institute in Brest, to track the protected Grey Seals as they swim to Britain. The device has shown that they can make the 186-miles journey in less than 48 hours and can swim at speeds up to 18ft per second. One report recorded that a year-old female Grey Seal, known to have a particular fondness for estuaries, swam to the mouth of the Thames, then headed back across the Channel making a brief stopover between Jersey and Guernsey, before returning to the point where the River Aulne meets the sea in Finistère. The monitoring of Grey Seals finds them crossing from Brittany to destinations as far apart as the Thames estuary, the Isles of Scilly and off the Coast of Cornwall, where they are known to mix with their British cousins for a couple of weeks; also there are regular observations around Les Minquiers. This research caused Susan Bell, a newspaper reporter in Paris, to caption an article: 'French seals seek love in British waters – A protected species of grey seals from Brittany is making excursions across the Channel to frolic with its British cousins and perhaps enjoy a holiday romance!'

Some fortunate boaters might find themselves accompanied by Grey Seals, which live within the reef, or by small pods of Bottle-nosed Dolphins (Tursiops truncatus), which often come inside the reef at high tide. A recent study by the Marie Louis of the Université de Poitiers suggests that the Minquiers Bottle-nosed Dolphins are part of a pod of approximately 237 individuals that permanently live in the Golfe Normano-Breton and which prefer to spend the winter and spring in the area immediately to the east and south of the reef. An estimated population of 185 of these live around Les Minquiers (P. Chambers, pers. comm. & Jouault, 2010).

In 2011 it was reported that a massive research project, conducted over the past few years by the Groupe d'Etude des Cétacés du Cotentin based in Cherbourg, has been undertaking an identification project by photographing dolphins' dorsal fins. And their latest work has involved recording the creatures using hydrophones – underwater microphones – laid at the bottom of the sea. These findings are being shared with Société Jersiaise marine biology section which is chaired by Nick Jouault. The dolphins being studied at the Minquiers are thought to be one of the largest single groups in Europe (Chiang, 20011).

Low tide at the Minquiers brings a dramatic change of scenery. The retreating sea reveals massive granite boulders, smooth and rounded by the action of the sea, which helps to protect La Maîtresse Île by dissipating some of the energy of the waves. Most spectacular of all, a huge curving golden beach of shell fragments is revealed. Low tides are known locally as 'ormering tides'. The Ormer represents a large Mediterranean cousin of the limpet, found here at the northern extreme of its habitat and under rocks exposed at low water. In April 1951, the States Tourism Committee invited to Jersey a number of international journalists, photographers and newsreel men not only to show them round the island but, as a highlight of the visit, to take them on an ormering trip to the Minquiers. Apart from the newsreel men and photographers taking 'shots' of an ormerer at work, they spent their time clambering over the rocks searching for Cormorants' and Shags' nests, of which they found a few containing eggs, the party's visit to La Maîtresse Île generated some excellent publicity for Jersey (JEP, 1951).

The Minquiers has always been a good place to fish for this delicacy;

for their relative inaccessibility has prevented the over- fishing that has occurred around other islands. However, the gathering of them is strictly controlled by law, and may take place only between 1 October and 30 April, and then only on the first day of each new moon or full moon and the three following days, and the shell must be a minimum of nine cms. measured across the broadest part of the shell. As well as the date restrictions on ormering there are strict rules on the means of gathering. Scuba diving for them is not allowed and anyone caught has their diving gear and the boat they were diving from confiscated. The wearing of any kind of face mask is not allowed which rules out snorkeling or even wearing swimming goggles (see Collyer, 2000; States of Jersey, 1994). However, as Le Maistre (2011) recently recorded: 'Fishing excursions to the Minquiers are particularly well supported when there is a good "ormering tide"'.

Another protective regulation states that 'parlour pots' are not allowed to be used around Les Minquiers. These are basically lobster pots from which the catch has no means of escape. When a pot has a fixed rigid neck, lobsters and crabs can find their way out once they've eaten the bait. This means that if the pot is left on the seabed for more than a couple of days the chances are it will be empty. 'Parlour pots' have a floppy neck or some other one-way entry system so that whatever goes into them remains there' (Daly, pers. com.). Also, Jersey Law prohibits building work to be carried out on either Les Minquiers or Les Écréhous during the spring and early summer, in order to avoid disturbing birds nesting during their breeding season (see States of Jersey,1994).

On the highest tide little more than a tenth of a hectare of Maîtresse Île is left above sea level. With such a small area for plants to grow it is not surprising that fewer than 20 plant species have been found on the reef and Sue Daly records:

'that any of these have survived at all is something of a miracle when, in the autumn and winter of 1972/73, weed killer [sodium chlorate] was sprayed over virtually all of the vegetation. This extraordinary act was part of an ill-conceived plan to cover the island in non-native conifers, supposedly in the name of conservation. As the reef belongs ultimately to the Crown, an appeal was made to Her Majesty's Receiver General, P.L. Crill, which successfully prevented the trees being planted.

However, the damage was already done. The vegetation there was the kind perfectly adapted to withstand the rigors of winter storms, salt spray and summer drought and it was wiped out in one stroke. With it went the ground cover and associated insects so vital to migrant birds in spring and autumn. Erosion followed where much of the precious soil was left bare' (after Daly, 2004).

The Joint Nature Conservation Advisory Body and Société Jersiaise Report 'To investigate the condition of the flora, fauna and topography of Maîtresse Île, and to make recommendations for its rehabilitation' (1973), as requested by the Receiver General, recorded that: 'The recent clearance and the uprooting of indigenous Tree Mallow plants show a wanton disregard for the natural history of the island, and are likely to have exterminated completely, or nearly so, several species of plants and animals, but even at this late stage, it is likely that if the island was left without further meddling for a few years, most of the species would re-establish themselves.' The report quoted from a letter from the Director of RSPB, Peter Conder, who wrote:

'...I think it would be absolutely nonsense to try and plant trees on Les Minquiers [...for] it is entirely against the principles of good land management and certainly very much against the policy with regard to introductions and reintroductions, which have been adopted by all reputable bodies in the United Kingdom...'

However, the 1973 Report concluded in a more diplomatic fashion by stating:

'We are sure that past clearances have been undertaken with complete unawareness of their potential harm and we do not doubt the good intentions of the putative tree-planters who have been led to believe that trees anywhere and everywhere are an unmitigated blessing. Finally, we hope that this report may be instrumental in leading to a reconciliation of the conflicting views.'

The recommendations submitted to the Receiver General, from the Joint Advisory Body and Société Jersiaise, resulted in:

'prohibiting the future extensive use of weed killer. Any large scale attempts at "tidying-up" the island, as the soil depends on the dead vegetation for its retention and replenishment. And any major disturbance of soil, as would be required for tree-planting' (after Le Sueur et al. 1973).

As Daly records:

The brightly-coloured Clown-faced Bug (*Pyrrhocoris apterus*), which has a striking red and black geometrical pattern on its back, and resembles an attractive beetle, was one of the early casualties of such unintended, but subsequent environmental vandalism. However, after the aforementioned conservation measures for the Maîtresse Île were put in place the vegetation has slowly returned and now Sea Beet (*Beta vulgaris*) and Tree Mallow are again abundant, although some of the original plant species are yet to return. The Cloud-faced Bug occurs in Jersey almost wherever there is a stand of its host plant Tree Mallow, and is particularly common on Les Écréhous, the Icho Rocks and now again on Les Minquiers (see Daly, 2004).

In order to highlight some of the flora which are known to have grown on three of Jersey's offshore reefs (Les Écréhous to the north-east, Les Minquiers to the south, and the Paternosters to the north-west) on 5 August 2003 Jersey Post issued a 'First Day Cover' colour envelope, designed by Colleen Corlett, with five stamps depicting various flora found on the reefs. Two of these showed the small cottages on La Maîtresse Île, with one having an illustration of Thrift (*Armeria maritima*), and the second with an illustration of a Smooth Sow-thistle (*Sonchus oleraceus*) which are both found on La Maîtresse Île (*Jersey Post*, 2003).

The Ramsar International Convention was signed in Ramsar in Iran on 2 February 1971, and represents a treaty designed to stop the encroachment on to wetlands and promote a greater understanding of the conservation importance of such areas. In November 2000 the south-east coast of Jersey was designated a wetland of international importance under the Ramsar Convention. In 2003, a consultant process started in order to give the Écréhous and Minquiers similar recognition as unique marine habitats and designation under this international convention.

At this time, the States of Jersey's Sea Fisheries and Marine Resources Advisory Panel gave its unanimous support to the designation process for these two offshore reefs also to be recognised under the Ramsar Convention, as did the Jersey Fisherman's Association (see Heuston, 2003).

At the beginning of February 2004 Jersey held a Wetlands Week to mark the 32nd anniversary of the Ramsar Convention. It was recorded that: 'The week included a series of walks and talks, and highlighted how over 32 square kilometres between St Helier and Gorey Pier in Jersey are covered by the convention and recognised internationally as significant ecological sites; other areas of interest around the island, as St Ouen's Bay, the Écréhous and the Minquiers, were also hoped to be soon incorporated within the Ramsar designation as well' (see Queree, 2004). At the same time, Andrew Syvret, Chairman of the Marine Biology Section of the Société Jersiaise, spoke on the subject of 'Ramsar - What Next?' as part of the ongoing dialogue concerning these offshore reefs as Ramsar wetlands of international importance, (Syvret, 2005 a) In March 2004 Mike Stentiford recorded:

'How the two reefs - Les Écréhous and Les Minquiers - are of enormously diverse fascination, not only to the avid fisherman or temporary recluse, but also for anyone with an up-tempo passion for wildlife and history. Of particular wildlife interest on Les Écréhous, for instance, is the annual tenure of Jersey's largest breeding colony of common terns ... Additionally, good numbers of cormorants ignore Jersey's rugged coastal mainland, preferring instead to bring up their families in the comparative peace and quiet these distant reefs provide. The seasonal bed-and-breakfast derived from the offshore vegetation also provides a critical lifeline to countless numbers of small birds during each spring and autumn migration. Added to this list of natural credits are visits by grey seals that commandeer the reefs for periods of leisurely solitude. It does seem to me that whatever our pursuits and interests, an overriding pride in these special offshore islets is obvious. Harnessing this pride with the protective obligation that Ramsar designation would bring is surely the correct and responsible step for our authorities to take.'

On 2 February 2005, World Wetlands Day brought good news for three of Jersey's most spectacular pristine and fragile maritime areas:

'The offshore reefs of Les Minquiers, Les Écréhous combined with Les Dirouilles and Les Pierres de Lecq or "Paternosters" were added to the United Nations Ramsar List of Wetlands of International Importance.' In addition to the already existing South East Coast Ramsar site, the Bailiwick of Jersey now boasts a total of some 18,748 hectares (72.4 square miles) of globally recognised inter-tidal and shallow marine habitats' (Syvret, 2005b).

By the designation of these offshore reefs as Ramsar sites of international importance, it was recorded:

'Such positive initiatives prove that environmental commitments can be achieved by the States of Jersey, and indeed by all of us, should we fully engage and operate our hearts, our minds and our green credentials. Full support for financial resources and, even more importantly, agreement to offer full protection to all those magical natural attributes Jersey should be so proud to have, must be uppermost on the list of priorities of every States member and government official' (Stentiford, 2008).

An integrated coastal zone management strategy 'Making the Most of Jersey's Coast' was lodged for States debate on 15 July 2008. The strategy document highlighted: 'The coast and seas around Jersey are an integral part of Island life. It is therefore essential that the coast is protected and managed so that it can continue to be enjoyed by generations to come'. Policy options in the strategy brought together a list of outputs and outcomes from which the Minister will be able to keep an eye on the key elements that will help to secure the long -term sustainability of our coasts and sea. These include Management Plans for Les Écréhous and Les Minquiers agreed by shareholders, with an expected outcome of improved management of the coastal and marine environments in Les Écréhous and Les Minquiers Ramsar areas (States of Jersey, 2008).

The Planning & Environment Committee of the States of Jersey designated the Minquiers as a 'Site of Special Interest', intended to protect the islets from insensitive development. The stated objectives of the of the States' Fisheries & Marine Resources Advisory Panel are to: 'Ensure sustainable use of the marine resources of the Bailiwick; maximise overall benefits to the people of Jersey; and to protect the marine environment.' In 2009, St Ouen's Bay / Les Mielles represented the fifth Ramsar protected area to be listed for future designation which, with previous Ramsar designations, is recorded under this International Convention as UK 23301-23305.

The three-day forum for the 'Future of St Aubin', held at the St. Brelade Parish Hall in January 2008, resulted in eight Action Group reports being presented to a subsequent meeting of the Parish Assembly. The Action Group on Ramsar Status was chaired by Michael Marett-Crosby, and the report recommended Ramsar status for the west end of St Aubin's Bay to be included as a priority in the 'Island Plan Review 2008' documentation. For the Action Group had gone deeply into the threats to this important coastal ecosystem, and the supporting conservation evidence within Jersey and internationally, as well as the procedure for designating this part of St Aubin's Bay as a wetland area of international importance. The Group also recognised that given the whole of the Golfe Normo-Breten shares a single tidal system with Les Minquiers and Les Écréhous it could form an important extension to these existing offshore Ramsar Sites. After the eight Action Group reports had been tabled and accepted by the Parish Assembly it had been the forum's understanding that they would be then forwarded by the Connétable, on behalf of the residents and community of St Aubin, to the relative governmental departments of the States of Jersey for their consideration and possible action (see Parish of St Brelade, 2009)

The international importance of Jersey's coastal waters is recognised by the fact that almost 190 square kilometres of inter-tidal habitat, spread across Jersey's south-east coast and off-shore reefs, including Les Minquiers and Les Écréhous, are now designated as wetlands of international importance under the Ramsar Convention. In January 2010 the convention included 159 member countries with a total 1,886 Ramsar sites recording 185 million hectares of protected wetland areas which cover all geographical regions of the planet. In March 2010 it was reported that Jersey's four most environmentally sensitive sites now had an authority to watch over them. The chairman of the new Ramsar Management Authority, Assistant Planning and Environment Minister, Rob Duhamel, said that this watchdog group would "have some teeth" because it was being led by a politician, and could recommend that legislation go forward to the States for consideration if it is thought to be necessary (after Ramsar, 2010; JEP, 2010).

In February 2011, the States of Jersey Department of Planning and Environment (P & E) published an important 38 -page policy document: 'Jersey's South East Coast Ramsar Management Plan' to coincide with the 2 February 2011 World Wetlands Day [1971-2011 40th Anniversary of the Ramsar Convention]. The Plan was written by the Ramsar Management Authority which was set up by Senator Freddie Cohen, Minister for P & E and is made up of government and parish representatives, the Receiver General and various non-governmental organisations, which included John Le Gresley representing hut owners on Les Minquiers. After the plan's publication the Assistant Environment Minister and Ramsar Management Authority chairman, Deputy Rob Duhamel, said: 'This plan is a major step forward in the responsible and integrated management of the south-east coast of Jersey ...We now need to work on achieving the aims it sets out and deliver the benefits for all Islanders and future generations' (see www.gov.je see States of Jersey, 2011; Hutchinson, 2011).

The Management Plan for Jersey's south-east coast highlights the main environmental and ecological threats to attaining the vision of the Ramsar Management Authority. They are: habitat decline due to land reclamation, pollution and climate change; conflict of use from fishing, aquaculture and recreation; invasive species; and limited jurisdiction. The Plan also recognises the importance of integrated environmental management, the protection of species and habitats, and the restoration of degraded habitats in the designated Ramsar area (3,210 hectares); and their conservation for future generations (after States of Jersey, 2011).

Paul Chambers (2010) summarised the current status and potential problems that could face these offshore reefs as follows:

'At present there is little major threat to marine life at Les Minquiers as the reef is not heavily frequented by leisure boating and is at present not commercially over-fished. Those that do visit invariably follow the marks to and from Maîtresse Île with few people daring to venture within the central reef. The number of shipwrecks around the reef attest to their danger to shipping and there is always the risk that an oil tanker (or some similarly hazardous cargo) might be spilled onto the reef...' (see Chapter four 'Reefs and Wrecks').

'The effects of disease and illegal fishing for ormers remains a potential problem. This especially true of the Vibrio virus which was estimated to have killed 90% of the ormer stock at Les Minquiers in the late 1990s. The stock has not fully recovered from this tragedy and it is speculated that the gyratory current which surrounds the reef may be preventing the recruitment of fresh larvae from the French and Jersey coasts. Strangely, this argument does not seem to apply to some invasive non-native species, such as the Pacific Oyster (Crassostea gigas) and the seaweed japweed (Sargassum muticum) both of which are now common across the reef. Jersey and Guernsey are receiving new invasive species on an annual basis; many of these seem to have been transported by visiting boats. Even with low levels of sea going traffic, it seems likely that the reef will become home to some these species in years to come.'

'The establishment of Les Minquiers as a Ramsar site gives cause for hope that the reef's biodiversity and ecology will at some point be evaluated and any necessary actions taken to minimise any deliberate and thoughtless destruction to its plants and animals. This applies not just to the wonderful marine life, but also to the few hardy terrestrial species that cling to those bits of rock which are not subject to routine inundation by the sea' (after P. Chambers, pers. comm.).

The establishment of a Coastal National Park took a step forward in the new draft Island Plan which highlights key policies and proposals and their possible impact on island planning over the next ten years. New policies apply to a marine zone, extending from the high tide mark to the edge of the island's territorial waters and to a proposed Coastal National Park. The proposed extent of the park would take in not only the lowland adjacent to St Ouen's Bay, north and south coast cliffs and the offshore reefs of the Paternosters, the Dirouilles, Les Écréhous and Les Minquiers, but also substantial parts of valleys leading to Gréve de Lecq and near St Catherine (after Shipley, 2011).

In June 2012 it was recorded that plans to build 33-foot weather masts on four sensitive coastal sites – including the offshore reefs of the Écréhous and the Minquiers – The £50,000 project would record data such as wind speed during a three-year trial to ascertain various information, including whether Jersey could use tidal and wind power as energy sources in the future. Yannick Fillieul, business manager of the Environment department, stated that the 2011 Island Plan, which sites planning policy, makes reference to investigating the long-term possibility of tidal and wind power as a source of energy, and that the information gathered from the masts would help Jersey understand if this was possible. However, she went on to mention that it was 'very unlikely' that the Écréhous would be used as a wind farm, and had added that her department was in discussion with users of the Minquiers to see where best to position the masts (after Hutchison, 2012).

In January 2013, the JEP reported that 'Plans for masts on the Minquiers and Écréhous are scrapped', after serious concerns were raised by islanders. Instead the Environment department submitted plans for the masts which could be placed on outlying rocks further out to sea, one 3 km away from the main Écréhous island, and the other 6 km away from the Minquiers. Although the project was yet to get the go-ahead from planning (after Stephenson, 2013).

However, with regards to the proposal to put ten metre masts on the offshore reefs to measure wind, Michael Dryden, Chairman of the Ornithology Section of Société Jersiaise wrote to the JEP to point out that the main source of danger to birds comes, not surprisingly, from the height of the masts in such exposed locations and, in particular, from the steel guy lines needed to anchor the masts against the wind which is being measured. In high winds and/or poor visibility, there will be collisions with these lines by large birds which will, in all likelihood, damage wings and therefore condemn them to a slow death. Mitigation measures proposed so far, would not help much in this respect.

Ornithologist, Dr Glyn Young, also expressed his concerns with regards to the risk of birds, such as shags and cormorants, flying into the wires used to support the structures and equipment from high winds, Although he said that he had been glad to see that the welfare of bird species in the area was being taken into account by planners when they Michael Dryden and he had been assured that there would be no routine visits to the weather masts, and for the data from them would be relayed back to Jersey through mobile phone technology. Other visits to the equipment would only occur outside of the breeding season. Dr Young emphasized the importance for such a degree of sensitivity to continue throughout the project.

In September 2013 it was reported the 'Offshore reef weather station up and running', after more scientific equipment has been installed on Jersey's offshore reefs after bird breeding colonies had delayed the operation this summer. A weather station has been put on Les Maisons, which is 6-km from the Minquiers, after workers had to wait for nesting birds to leave their nests and for the young to fledge. Jersey Met hopes that the data can be used to future planning involving renewable energy resources and that they van now start work on setting up the transmissions of this important data (after Chiang, 2013).

As Paul Chambers, Marine and Coastal manager at the Environmental Department recorded (2017):

'In September 2017, the States of Jersey approved two protected areas. However, these new zones were far from straight forward. Both the Écréhous and the Minquiers are located outside of Jerseys exclusive three-mile limit fishing zones and, although they lie within the island's territorial waters, commercial fishing there is managed via an agreement between the UK and France, known as the Bay of Granville Agreement. Under this any change to fishing practices requires the approval of the Jersey and French authorities and, for pragmatic and political reasons, it took nearly five years for Jersey to get consensus with the French authorities for the new protected areas.'

'A century ago, it was reported that French scientists (and others since) suggested that Jersey's offshore reefs were important nursery grounds for fish and sheefish species. Because situated offshore and washed daily by 12-metres tides producing a complex and productive food chain and the rocks, seagrass and seaweed, provide sheltered breeding conditions for fish, scallops, lobster and crabs. This nursey aspect of the reefs was an important argument in favour of protecting the seabed. This, and the island's international requirement to conserve seagrass and maerl habitats, were central to a request for protection that Jersey put forward to the Bay of Grouville Agreement in 2013.'

'The case was presented by the Environment Department and Jersey Fishermen Association, with full support of the Minister of the Environment, Steve Luce, and the zone around the Minquiers was agreed quickly; largely because the reef is little used for dredging and trawling. The consensus reached in Granville in February, resulted in the protected zones becoming legal from 27 September 2017. News of the two new protected marine areas received praise from local and international organisations (Chambers, 2017).

In April 2020 a total ban on catching the Atlantic bluefin tuna *Thunnus thynnus* in Jersey waters was agreed in order to help preserve the species. Environmental Minister, John Young, signed a ministerial decision completely blocking access to the species for any vessel later in 2020. Under the change to the Conservation of Wildlife (Jersey) Law 2000 any catching of the tuna in the area will be banned. Deputy Young said, 'by prohibiting the landing of Atlantic bluefin tuna, Jersey is joining international efforts to protect the stock for generations to come' (Maquire,2020).

Chapter 4

Reefs and Wrecks

In early times Chausey and the reefs of Les Minquiers must have represented fearsome barriers to a coastal navigation that relied on oar or sail, and though they provided a regular harvest of wreckage. Brannan (2003) records that a certain amount of historical evidence concerning shipwrecks and drowning. In 1615-1617 Captain Collas Grandin and his crew came before the Manorial Court of Noirmont in connection with the salvaging of the wreckage of a ship from the Minquiers. It is also interesting to note that the archives held by the Société Jersiaise include documents relating to a 1692 dispute between the Dame of Samarès and the King, regarding the ownership of wreckage in the Minquiers; and a summons to the Attorney General and Receiver to appear before the Privy Council concerning Deborah Dumeresque's appeal in her claim (Société Jersiaise, 1692).

When sailing on the SS Crusoe of Jersey to St Malo in January 1748, Jean Hamon was the sole survivor of a wreck on the Minquiers. Hamon recorded in his journal how he had eventually scrambled onto La Maîtresse Île where there was no shelter of any kind; no fishermen's huts had been built at that time, and he was forced to rob the bodies of his drowned friends to keep warm. He collected fresh drinking water in empty Ormer shells and ate raw fish and the flesh of seabirds that he managed to kill with rocks. After a month of privation, Hamon was eventually seen and rescued by a passing French ship (after Brannan, 2003).

HMS *Grappler*, a British gun brig cutter, built in 1797, armed with two 24-pounders and 18-pounder carronades, ran aground and was wrecked on Plateau des Minquiers on 31 December 1803. On 23 December the Grappler, under the orders of Admiral Sir James Samarez, had sailed from Guernsey to take four French prisoners, two old men and two women, to Granville to release them. The same evening a gale forced Lieutenant Thomas to take shelter in a difficult anchorage under the Île de Maître, one of the isles of Chausey. The gale continued for a few days and it was not until the 30th that the weather moderated sufficiently to attempt a return to Guernsey, his prisoners having agreed to be left on the island if he provided them with a boat and provisions. Unfortunately, while getting under weigh a hawser parted and HMS *Grappler* drifted further to the north-west on to a half-tide rock on the Minquiers reef and broke in two when the tide dropped. There are contradictory reports as to whether the cutter was wrecked on one of the Chausey reefs, or close to La Maîtresse Île on the Minquiers reef. Although there are reliable reports that on the 31 December 1803, after an exchange of fire with the French, the Grappler was set alight and Lieutenant Thomas and his crew were taken prisoners by men under the command of Captain Epiron of the French Navy (after UK Hydrographic Service Report, update 2007).

In 1811 a case was brought before the Royal Court claiming the compensation due to a group of fishermen for their assistance to a ship wrecked on the Minquiers. And in 1817, a similar case was brought by the owners of the cutter SS *Ross*, who had gone to the Minquiers to cut vraic and helped recover part of the cargo of the Jersey vessel *Minerve* (Falle, 2001). On 31 July 1816 the newly built 39-ton French sloop MV *Hirondelle*, under Captain Fulliade Pierre, foundered to the north of Les Minquiers on her way to St Malo (UK Hydrographic Service Report, update 2009).

In September 1850, a passenger paddle steamer SS Polka was on passage to St Malo from Jersey, under the command of Captain Priaulx, when it sprang a leak. As the ship was not far from the Minquiers reef Captain Priaulx was able to steer his ship to the La Maîtresse Île, and successfully to transfer his passengers to the islet before the ship sank in deep water. Sadly, the story did not end there, for eight days later on 24 September 1850 Captain Priaulx now in command of the 150 ton SS *Superb*, while steaming from St Malo to St Helier, foolishly was prevailed upon by his passengers to visit the scene of the first wreck. To the misfortune of all concerned the Superb, travelling at full speed, struck the rock (Grune de Turbot) near the Maîtresse Île and quickly foundered. In the ensuing panic 17 passengers were tragically drowned. Fortunately there were fishermen nearby who came to the rescue. One of the fishermen, Thomas Hamon, had observed floating wreckage and was able to save three lives and recovered one body. By 1873 the fishing fleet at La Rocque numbered about thirty boats, with the majority of them taking advantage of the fish resources surrounding the Minquiers. On 19 February 1887 the MV *Mera* was reported missing within the latitude and longitude of the reefs of Les Minquiers (see Falle, 2001; Jouault, 2001; UK Hydrographic Service Reports, 1999-2010).

In 1891, after the wreck of a French cutter was found with the loss of 51 lives, a French lightship was positioned off the plateau, just outside the three-mile territorial limit, but when France replaced the lightship with eight navigation buoys encircling the Minquiers, French fishermen claimed territorial rights (see Jefferson, 1985; & Chapter 5 of this publication). On 7 November 1902 the French 300 ton wooden-hulled Brigantine SV *Gabrielle*, carrying a cargo of fish from St Malo to Bordeaux, was considered to have foundered in the region of Les Minquiers reefs (UK Hydrographic Service Report, 1999).

Seaplane Cloud of Iona *which crashed with total loss of life on 31 July 1936.*
Photograph Ref: SJPA/009887, 1936 © Société Jersiaise

The last hours of the flying boat the *Cloud of Iona* has an interesting link with the Minquiers (see plate 15). On 31 July 1936, the *Cloud of Iona*, owned by Jersey Airways and carrying eight passengers and two crew disappeared in bad weather on a flight from Guernsey to Jersey. The Iona carried no wireless but did have Verey lights and distress flares, and enough fuel to last nearly two hours' flying time. When the Jersey Company's traffic manager realised that the flying boat was overdue, with the weather worsening and no signs of the plane, the Guernsey lifeboat was launched and followed the same route that the Iona had planned to take. Soon after this, the French Consul in Jersey alerted all the coastal towns between Cherbourg and Cartaret. With visibility down to 30 yards the States of Jersey tug the *Duke of Normandy*, steamed out into the storm, and about the same time the cargo ship, MV *Roebuck*, set out from Guernsey to help in the search. The following morning a 16-seater express owned by Jersey Airways Ltd took off from Southampton and flew the length of the French coast at a height of 50ft. And on the same morning an RAF flying boat joined the search (after Cook, 1967).

The day after the disappearance of *Iona* a notice was placed in the local newspaper asking any person who had seen or heard any aircraft over the Island, or over or on the sea, to inform the named authorities. In spite of varied reports it was not until 24 hrs later that a French rescue craft sighted debris nine miles south-west of Corbière. Then came a most important discovery, wreckage was seen at the Minquiers and an expedition under H.G. Benest, the Lloyd's Agent, recovered the plane's emergency door three miles from the Maîtresse Île. They also found portions of fabric, plywood and air cushions from the pilot's cabin two miles west of the Pipettes rocks. The fate of the *Iona* was known but its last, catastrophic landing place remained a mystery. The task of piecing together the sea-plane's last hours was puzzling enough, for strong tides and high winds had washed up wreckage from the plane so many miles apart. However, almost a fortnight after the *Iona*'s final flight, two Jersey fishermen, Winter Gallichan and George Marie, came across the aircraft's engines and its smashed hull wedged between the rocks at Les Pipettes near the Minquiers. The rocks were only visible at low tide, but it was obvious that this was the *Cloud of Iona*'s last resting place (Cook, 1967; see also JEP 1936 & 1937).

In July 2019, on the 80th Anniversary of the tragedy, the JEP published under 'Historic Headlines' two lead articles 'The seaplane that was lost at sea with all on board', and 'A flight lost at sea'. These included 12 b. & w. images and made reference to the pilot of the seaplane, William Halmshaw, an ex RAF officer who was well known in both islands and had considerable experience with flying boats. Shortly after the accident his wife said that he would never take risks and she thought he had probably landed in France. The engineer, Mr Sotinel, a St Aubin resident who had previously served in the French Air Force (Heath, 2019).

On 22 August 1944 it was reported that the German steamer, MV *Fink*, had foundered on the Minquiers reefs, although due to the ship's loss occurring during the latter part of World War II, no further information at the time of the wreck was recorded (UK Hydrographic Service Report, 2008).

Just after 3 a.m. on 15 December 1953 an SOS message was received by the Jersey Harbour office from the 662-ton Bristol trawler MV *Brockley Combe*: 'Have struck rock on Minquiers, ship making water in engine room, need immediate assistance' (see Plates 16 & 17). The St Helier lifeboat, MV *Elizabeth Rippon*, and the States Tender, the Duchess of Normandy, put to sea with all possible speed. The *Brockley Combe*, built in Bristol in 1938, was on passage from Guernsey to Jersey with a part general cargo when it hit the reef. Subsequent messages from the lifeboat showed that the stranded ship had been located in the middle of a mass of rocks, and that the lifeboat was standing by. Also, that prior to the *Brockley Combe* breaking her back on the rocks and becoming a total loss the crew of eleven, in very poor visibility, had been safely taken off the merchantman and were being brought back to St Helier (JEP, 1953; UK Hydrographic Service Report, 2008).

In August 1956 some mystery surrounded an incident that had occurred at the Minquiers, when two French fishermen were rescued by a yachtsman from Cancale two days after their boat, a ketch, had broken adrift from its moorings at La Maîtresse Île and had drifted on to rocks near Lessay, on the French coast. Mr Gordon Coom of the South Pier Shipyard, whose family owned one of the huts (the Old Hospital) on La Maîtresse Île, had a few days previously been fishing on his boat, Happy Returns and seen the French fishing boat anchored in the bay. When he met the fishermen on the reef they had told him that they had run out of food, so he had given them bread, tea and tobacco. After their ketch had

Bristol coaster Brockley Combe *wrecked on Les Minquiers December 1953.*
Photograph Ref: SJPA/014348, 1953 © Société Jersiaise

The skeleton of the Bristol coaster Brockley Combe *after over fifty years of tidal surges and gales at Les Minquiers, 2010*

disappeared from its moorings and prior to their being rescued, it was brought to light by the captain of the States' tug, Mr G. Marie, that the hut owned by the Coom family had been broken into and showed signs of recent habitation. There was a quantity of seabird eggs, fresh boiled prawns and ormers, and the floor of the hut was littered with empty food tins. A mattress and pillow had been laid on the floor. The names of the fishermen were yet to be known (JEP, 1956).

Mauger (1978) wrote about his voyage to the Minquiers: 'The most dangerous group of rocks in the Channel Islands, a ships' graveyard where heavy losses have occurred over the centuries'. And he made reference to how the reef leapt into the limelight when Hammond Innes published his exciting novel, *The Wreck of the Mary Deare* (1956), which is set in the Minquiers. This was followed by the film version starring Gary Cooper and Charlton Heston. Mauger reflected how Hammond Innes always wrote about man's battle against the forces of nature, and in the Mary Deare he chose one of the most dramatic and fearsome locations it is possible to imagine. Mauger's article concluded: 'In many impressions gained at the reef one is outstanding – How remarkable it is that rocks are in just the right place to act as navigational marks, Clearly the result of some cataclysmic upheaval when earth was formed.'

As recently as the summer of 1968 a St Helier fisherman, Mr Terry de Ste Croix, while ormering on the Minquiers found a bronze 16th century Spanish cannon. The cannon, a comparatively rare breech-loader weighed 2 ¾ cwt. and was situated deep in a crevice in the rocks but he managed to free it 'after a long struggle' and bring it back to Jersey. He reported his find to the Harbours Office and to the Société Jersiaise, and photographs and notes were taken by the Société of the cannon, which was found to be in excellent condition. The details were forwarded to the Armoury at the Tower of London, Woolwich Arsenal, and the Royal Military Academy, Sandhurst. Their replies agreed that it was a Spanish cannon dated between between 1555 and 1600. It was valued at approximately £100, although a London antique dealer considered the cannon to be worth £300 to £350 (JEP 1969).

Another discovery was made on the Minquiers, Here, in 1985/86 Tony Titterington found a M343, a German anti-submarine vessel that had been en route from Guernsey after D Day in 1944, when she was attacked by allied MTBs. Lying 100 feet down at low tide, she too was carrying a submarine gun, this one being a 10.5 cms model. Tony, who had been diving with his friend, John Blashford-Snell, found that the vessel had a severely damaged amidships by the explosion of a depth charge, probably detonated by the MTBs canon fire. They found the gun was loaded and, after quite a struggle, managed to open the breach and extract a complete cartridge and shell. As Blashford-Snell records:

'At that time, a number of Jersey Police were being trained as divers by Tony and they watched in awe as the live shell was lifted to the surface in a net attached to an airbag. Once at the surface Tony placed the shell on the deck of his skiff where the police friends were seated. Suddenly the shell began to 'hiss' and, in a trice, the gallant policemen were over the side into the sea. There was a loud 'poop' and the shell popped out of its case and shot across the deck of the skiff. Roaring with laughter Tony teased the policemen. "It's only the change of pressure in the shell case that cause this" he explained, as some very sheepish officers climbed back aboard.

Carrying the live shell gingerly over the shingle and up a low cliff to his home [in Jersey], Tony deposited it in a retort clamp outside his workshop. Visitors staying at Tony's Mother's hotel often caused to eye this and noticed a cup of plasticine over the fuse head. "Oh, that's to hold the Nitric and Hydrochloric acid" Tony explained "to eat away the fuse". The guests quickly dispersed. Later, the salvagers (on the Minquiers] found a stock of depth charges and torpedoes in the hold and they are probably still there!' (Blashford-Snell, 2020)

Numerous articles have highlighted the sinister navigational dangers of the reefs of Les Minquiers be they notorious reefs, wreck-strewn sandbanks or vicious tide races (see Cumberlidge, 1990; Lasaygues, 2006). Lemprière-Robin & Falle (1986) record how: 'The Minquiers reef tends to be out of bounds to the inexperienced yachtsman. Navigation can be hazardous among these rocks amid the rise and fall of the tides and the fierce currents. The easiest passage into the anchorage at the Maîtresse Île is from the south-west. Those who prefer to live dangerously enter from the north through the formidable Gauliot Passage where great volumes of water are forced through a narrow gully lined with jagged rocks and the helmsman is wise not to let his attention wander.'

However, as Peter Carnegie points out, when making a first acquaintance of Plateau des Minquiers the timing of a visit and selecting an anchorage is best over high water neaps, in settled weather with at least three nautical miles visibility for identifying marks: 'Aim to be at the North entrance around half flood so entry is made on a rising tide with decreasing current. Piloting in at an average speed of four knots and allow about half an hour to reach the Pool [anchorage by slipway]. At least three hours remain to take in some extraordinary scenery before returning on the ebb'... 'In view of the dangers in the approach to the anchorage NW of Grande Gauliot it should not be attempted without local knowledge on board. From the highest point on the north end of Maîtresse Île it is possible, on a clear day, to pick out the spire of St Malo Cathedral' (Carnegie, 2006).

Accidents that have occurred on the Minquiers and requiring assistance, have always been quickly responded to by the RNLI, as well as more recently by the Jersey Lifeboat Association's all-weather vessels:

'For in July 2015, the RNLI boat collected an injured hut owner from La Maîtresse Île (see colour plate). And on 31 May 2020, three people were rescued after their sailing yacht had broke its moorings and sank. The RNLI's in-shore rib, and the Jersey Lifeboat Association's all-weather vessel were called late in the evening, after receiving a report that the yacht was taking on water just to the south of Les Minquiers; a French Naval helicopter was also dispatched. The three crew members managed to scramble onto the rocks of the main island and were picked-up by the inshore lifeboat and transferred onto JLA's vessel, and brought back to St. Helier (after Heath, 2020).'

As Falle (2001) records:

'Those who are lucky to have stayed on the reef for the full cycle of a neap or spring tide; to watch in solitude under a great and ever changing sky the plummeting of gannets and terns, and the graceful passage of dolphins; to be eye to eye with great black-backed gulls or the occasional seal; and all this in the company of friends; is to feel as near to peace with Nature as anywhere in the Bailiwick of Jersey.'

Chapter 5

French Claims to Les Minquiers

Until 1929 no one doubted that the Minquiers belonged to Jersey. Acts of the Seigeurial court of Noirmont in 1615 and 1617 show that the Crown claimed all wreckage on this reef. In 1692, and as previously noted (see Chapter One: Geography and History), the Dame of Samarès disputed this claim, asserting that the rocks formed parts of her fief; but she lost her case. La Maîtresse Île has always been a resort for Jersey fishermen, although in 1929 the matter of sovereignty was brought to a head when a Paris banker Henri Leroux, obtained from the Land Registry at Coutances (Normandy) a lease of part of La Maîtresse, and began to build himself a house. Strong protests were raised, and while Jersey fishermen hoisted the Union Jack, Monsieur and Madame Leroux cavorted round the reef in blue, white and red outfits masquerading as the Tricolour (Balleine, 1970 [as revised by Joan Stevens]; and see James, 2007).

Between the 9 and 12 July (1929) the story was reported in the *London Evening Standard*, the *Morning Post*, the *Daily Sketch*, the *Daily Mirror*, and in *Le Petit Parisien*. The *Daily Mirror* reported the event under the bold headlines 'Comedy of the Island War - Twelve men and a dog defend the Union Jack'. Soon, many of the national papers of both the UK and France had taken up the story. On Monday 15 July the *Jersey Evening Post* reported that their representative had visited the Minquiers and met Monsieur Leroux, who was quite genial and showed a document headed '*Bureau d'Enregistrement, Coutances*'. It stated that he had a nine year lease of a certain portion of land on Les Minquiers, granted by the French Ministry at Coutances. He therefore refused to desist from building merely on the protest of private individuals (Reed, 1989; Richardson, 1952).

West side of La Maîtresse Île
Helicopter photographer, 2008 © Julian Mallinson

Low spring tide, looking south from the north of La Maîtresse Île
Helicopter photographer, 2008 © Justin Simpson

West side of La Maîtresse Île, Les Minquiers, July 1999
Photograph, Peter Mourant © Jersey Evening Post

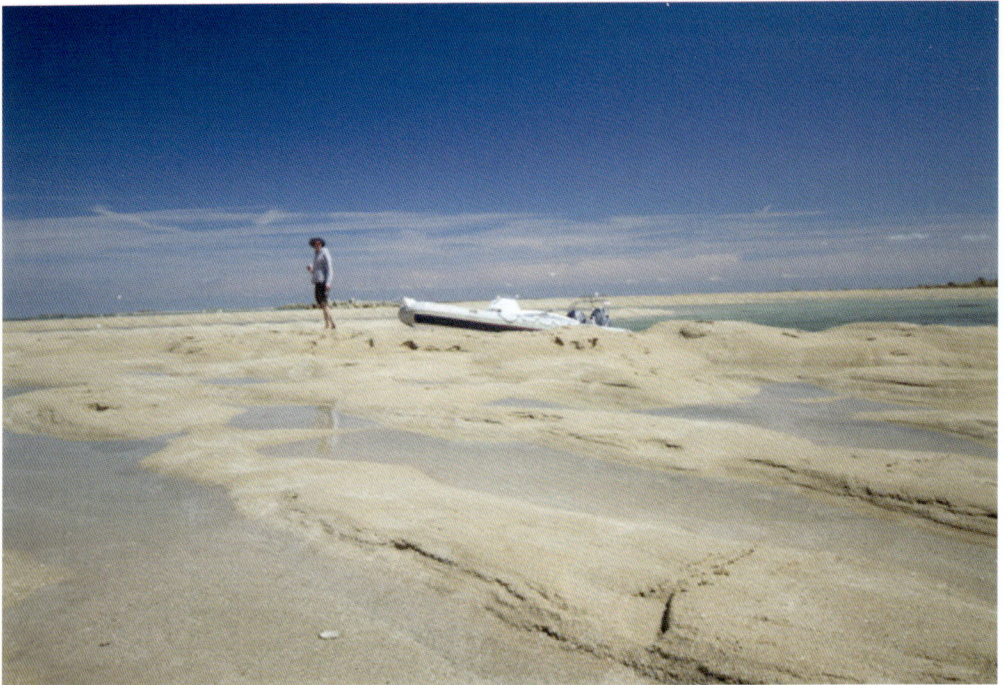

'Rib' on sand bar to north-west of La Maîtresse Île
Photograph July 2015 © Julian Mallinson

Rainbow over La Maîtresse Île taken at 08.30hr, 9 September on a 39 feet spring tide
Photograph 2010 © Julian Mallinson

The sand bar, as the tide ebbs, on the north-west of La Maîtresse Île, July 1999
Photographer Peter Mourant © Jersey Evening Post

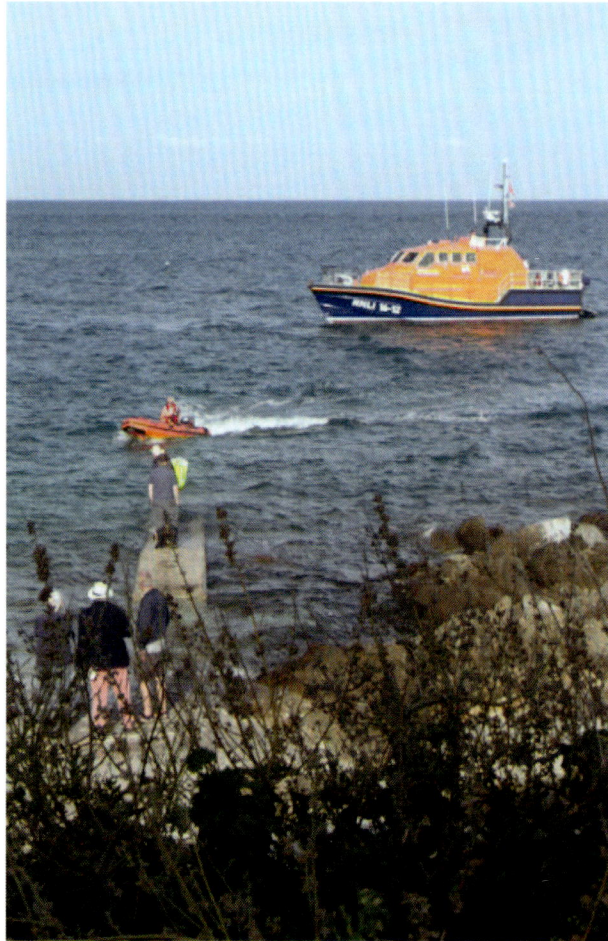

Left: RNLI boat, skippered by Andy Hibbs, collecting an iinjured hut owner from La Maîtresse Île
Photograph 31 July 2015 © Julian Mallinson

Below: Fishing at the time of the Summer Soltice
Photograph 2019 © Julian Mallinson

Visit of Lieutenant-Governor, Bailiff and States officials to La Maîtresse Île, a group of 29   Photograph Ref: SJPA/013110, 1896 © Société Jersiaise

British newspapers continued to report the affair, magnifying it out of all proportion and Mr Roy Bishop of the Morning Post wrote an article in humourous verse, headed 'War in the C.I.'. The French Consul in Jersey went to Paris and the Lieutenant Governor reported the matter to the proper quarter in London (Richardson, 1952). After many letters had passed between Whitehall and the Quai d'Orsay, the French Government asked him 'to consider as suspended the lease which permitted you to build on the Minquiers and to stop the building, adding, however, 'the question of the sovereignty of the Minquiers has not ceased to be contested between France and England' (Balleine, 1970; James, 2007).

In 1939 another Frenchman, Monsieur Marin Marie, took up the cudgels. Helped by a subscription raised among ardent nationalists he bought at Granville a prefabricated house. On the night of 10 June, he and his friends set off in 15 boats, and in 24 hours the house had been erected on Maîtresse Île. Then came the war. The Germans stationed an 'ack-ack' unit on the island and, to keep themselves warm, they stripped the cottages of every scrap of wood leaving nothing but bare walls (Balleine, 1970).

In 1953, the French claim to the sovereignty of Les Minquiers and Les Écréhous was submitted to the International Court of Justice (A judicial court of the United Nations) at The Hague. The French Government claimed that the Minquiers were a dependency of the Chausey group of islets, and that the Écréhous became their property by reason of the gift of Piere des Preaux to the Abbey of Val Richer in November 1203. M. Andre Gros, legal advisor to the French Foreign Office claimed prior right to the islets by title. At the hearing, he quoted from a 1897 dictionary of geography which gave the Minquiers as a dependency of Chausey. He maintained that the islets were in a French bay, a natural indentation of the French coast, and as such part of French territory, since, he claimed, the waters between them and the mainland of France were 'interior waters' (Smith, 1978).

Documentation, regarding the 1951 Agreement between the Government of the United Kingdom and Northern Ireland and the Government of the French Republic, concerning the Rights of Fishery in the areas of the Écréhous and the Minquiers (see HM Stationery Office, 1952) and the subsequent pleadings, oral arguments and documents relating to this case of sovereignty. These well cover the historical and legal arguments behind the 17 November 1953 ruling by the International Court of Justice that the islets and rocks of both Les Minquiers and Les Écréhous groups were part of the Channel Islands and did not belong to France, as the latter had claimed (see I.C.J, 1953). Also, documentation of the 17th century showing the reef to be part of the Fief of Noirmont (Jersey), proved to be an important proof of title. As Falle (2001) postulates: 'It seems reasonable to conclude that the Minquiers still form part of the fief of Noirmont.'

Rather than delving more into the historical and legal arguments behind the Court ruling, it is considered to be of particular interest to this publication to relate some of the oral and written pleadings of local people who had connection with the territories concerned; communications that have been so well recorded in an article by James Brannan (2003): 'All of such references were submitted to the Court as evidence that the islets and rocks, since 1066, had "always been part of the territory of the English Crown as dependencies of Jersey" and that France could not claim sovereignty on any such basis.' Brannan (2003) records:

'The British legal delegation notably included the Jersey Attorney General, Cecil Harrison, who injected much local colour into the proceedings. The hearings were actually attended by Écréhous resident [Captain] Bolitho and his wife [Roselle] née Lemprière, and the Court was told that the couple had "sailed [to The Hague] in their own boat from Jersey in order to attend". It was pointed out to the Judges that the "ancient Jersey family of Lemprière" [Rozel Manor] had owned property at Blane Île "for many, many years past" and that it would be "a rather remarkable situation" if they were to find that "all this time, without knowing it, they have been really living in France!'

Attorney General Harrison, stressed the importance of this ownership evidence and referred to 'the human question of people, actual people, who have properties on these islets – many of which they and their ancestors have held for very many years in the faith that that they are part of Jersey'. The case-file thus included a list of 20 named owners of Minguiers properties in 1888, and as for the Ecréhous, a letter of 1846 by W. Le Couteur, Viscount of Jersey, showed that Jersey people had houses there at least from the earliest part of the 19th century. References were also made to contracts made by Clement Gallichan to Josué Le Bailly in 1863 of a house in Trinity together with a property at the Écréhous, and in 1881 Lerrier Godfray selling a building on the Écréhous to Henry Bertram, which was subsequently sold a few years later to the Jersey Customs Authority.

During the oral proceedings of the Court Harrison received a letter from Jersey referring to a 'book of family history' once kept by the father of Fred Tocque, sail maker and former captain of the Duke of Normandy tug. The book significantly contained the following note: 'AD 1690 – Nicholas Tocque. Died and buried by the side of his house the south side of the Maître Île,' thus is providing much earlier evidence of a Jersey family's presence on the islet. Evidence was also produced of stone being quarried on the islets by Jerseymen as far back as 1623, as well as the great deal of quarrying that took place on the Maîtresse Île during a period of about 20 years, from the late 18th to the beginning of the 19th centuries, for the building of Fort Regent (after Brannan, 2003).

In 1953 a drama of historic importance to Jersey was played out on the international stage, when the International Court of Appeal ruled that the sovereignty over the islets belonged to Great Britain. This judgment did not affect earlier agreements, redefined in 1951, which gave equal rights of fishing to French and British nationals in the areas of the Écréhous and Minquiers i.e. 'between the limit of three miles calculated from low-water mark as reserved to French nationals by the 1839 Convention'(see HM Stationery Office, 1952). However, it was Professor E.C.S. Wade who transported the International Court of Justice back to the 10th century as he chartered the years to prove Britain's claim to sovereignty of the islets, when he said that France, even if some sort of abstract title could be postulated in favour of them, lost that title, and a valid title on the part of Britain had replaced it. Thus as age-old link was confirmed in areas which had seen much clandestine cooperation in earlier centuries (after Jamieson,1986; Smith, 1978).

As Richard Falle (2001) records:

'The results at The Hague, celebrated at a political level by lawyers as a leading case ...International Law was otherwise something of a Pyrrhic victory for Jersey. The French had only submitted to the jurisdiction of the International Court on the basis of an agreement with the United Kingdom government in 1951, that win or lose, they would continue to enjoy almost unrestricted access for their fishermen to both the Écréhous and the Minquiers. This permanent surrender of an historic possession accordingly made the question of sovereignty before The Hague almost academic. There are now very few Jersey fishermen on the Minquiers. The lobster fishery reckoned to be one of the richest off the coast of France, is now almost entirely lost to the Jersey economy'
(also see, Lemprière-Robin and Falle, 1986).

After the sovereignty of the reef was confirmed in 1953 in favour of Britain, the French removed all their buoys. Since then the States of Jersey, with the help of Trinity House, has maintained an excellent system of buoyage and marks around the reef. The location of the States buoy at The Pool anchorage at La Maîtresse Île is recorded as: 48°58′.16N

2°03´63 (after Carnegie, 2006).

In 2000, the States of Jersey asked the United Kingdom to execute an agreement with France to regulate fishing in the Bay of Granville. The agreement established the maritime boundary between the territorial waters of Jersey and France. It was given legal effect in Jersey on 4 July 2000. Unfortunately, as Richard Falle (2001) relates:

'The whole character of this agreement which was the result of a long and tortuous negotiation reflects the 1951 protocol which preceded the case before the International Court proceedings at The Hague. Although, in theory, the 2000 agreement has given rise to a common fishing regime and gives Jersey fishermen greater access to French waters than they have previously enjoyed, it is probable however, that it will make very little difference. However, in order to exercise an outward show of sovereignty there are annual visits: For the Crown, by the Lieutenant Governor; for the States of Jersey, by the Harbours and Airport Committee; and for the Parish of Grouville, by the Connétable and his Road and Rates Committee ...'

Chapter 6

National Flag Incidents

On 28 May 1945, nineteen days after the Channel Islands had been liberated from the German occupation, a ceremony took place on La Maîtresse Île when the Bailiff of Jersey, accompanied by an armed guard, the Commander of the British Army of Liberation, together with the Attorney-General and other dignitaries, re-established the Crown's tenure of Les Minquiers by hoisting the Union flag, and thereby formally retaking possession of the island in the name of Great Britain. At the same time, several French fishermen found on the island were asked to leave - it being pointed out to them that they were unlawfully present on British soil (after Butlin-Baker, 1981).

Just over two months later, the *Jersey Weekly Post* recalled a repeat of an incident in 1929, when the French tricolor was hoisted at the Minquiers by the French banker, during his construction of a house (see Chapter five: 'French Claims to Les Minquiers').When the States tug arrived at the Minquiers on Friday 3 August 1945, the crew were astonished to see the French flag flying from the top of what appeared to be a newly-erected flagpole. Going ashore they discovered that the pole had been recently erected, and a good job had been made of it, for it was cemented in, and the date of its erection - Aout, 1945 - neatly worked in the concrete The Union Jack, that had been hoisted by the Bailiff in May, was still flying bravely from another pole at the other end of the island. After this repeat hoisting of the French flag on the Minquiers, it was postulated that 'it seems likely that this group of rocks off Jersey's south coast is once more to become a subject of discussion' (see J.W.P., 1945).

In 1984 the French author, Jean Raspail, organised an invasion of La Maîtresse Île. On 3 June, *The Mail on Sunday* reported: 'Two years on and [after the Argentine invasion of the Falkland Islands] a new invasion scare for Whitehall ... Island raid by lunch party.' The article continued in a similar jocular fashion: 'Invaders, armed to the teeth with picnic baskets, had come from mainland France to this southerly point of the British Isles.' Two Jersey fishermen knew that the tiny outcrop had changed hands when they saw that the Union flag had been replaced by a strange red, white and green tricolor. One of the fishermen, Bob Le Marquand was quoted as saying: 'We were furious. We rushed back and pulled it down.' The reporter read the plaque left by the 15-strong landing party which declared their support for an obscure Frenchman who, they said, was the first king of Patagonia. And it declared the island Patagonian territory. 'The invaders toasted their cause with white wine before fleeing on a luxury cruiser. The invasion is being treated as a prank. But it underlines the bitterness towards Britain over sovereignty of the Minquiers - granted at the expense of France after the Second World War' (Dobbie, 1984). It later transpired that the invasion had been little more than a publicity stunt for Jean Raspail's new book (Petters, 1998).

La Maîtresse Île: Union flag being hoisted after the liberation
by 11th Battalion Hampshire Regiment on 28 May, 1945
Photograph Ref: SJPA/013111 © Société Jersiaise Photographer Raymond R. Marquis

However, fourteen years later Raspail was to repeat his invasion of sovereignty, for on 1 September 1998 the Paris correspondent of The Times reported: 'Channel islet seized by King of Patagonia - Invasion by the king of Patagonia' - A tiny British-owned island in the English Channel has been "invaded" by an eccentric French novelist in the name of King Orélie-Antoine I, the self-proclaimed monarch of Patagonia who died more than a century ago. Jean Raspail, a writer who styles himself consul-general of the non-existent Argentine kingdom, said a unit of 'Patagonian marines' under a retired English Rear Admiral had claimed sovereignty over the deserted Minquiers archipelago south of Jersey in retaliation of Britain's "occupation" of the Falklands. Raspail went on to declare: 'A light naval unit of the Patagonian fleet landed ... and hoisted the royal blue, white and green, flag in place of the British Flag, which can be honourably returned to Her British Majesty's Embassy in Paris' (Macintyre, 1998).

Macintyre's humorous report in The Times went on to record:

'The amphibious landing was carried out at dawn on Sunday and met no resistance since the only inhabitants at the time were a handful of surprised seagulls. In fact, the invasion passed unnoticed until the declaration of sovereignty was issued in the name of Orélie-Antoine yesterday. The British Ambassador in Paris was unable to comment on the invasion. Indeed, the Bank Holiday switchboard operator was rendered speechless by the news. M. Raspail said the fictional kingdom of Patagonia has more than 1,000 subjects to its name and boasts seven boats in its navy. He refused to identify the admiral in command of the royal fleet, pointing out as the kingdom's naval supremo, the retired British officer had felt it necessary to resign for 48 hours during the invasion in order to avoid a conflict of loyalties ... M. Raspail said he expected a British effort to retake the islands would be launched "within hours", but last night the blue, white and green of the royal house of Patagonia still fluttered over the disputed rocks, as Britain faced the first invasion of its Channel Island possessions since the Second World War.'

The so-called invasion party was witnessed by Advocate Peter Mourant and his wife Anne, who were with friends on a day trip to the Minquiers. As Leigh Petters (1998) reported:

'Advocate Mourant said that the "light naval unit" was little more than a French-registered speedboat, the number of which they have passed on to Jersey Police. The two men and two women who staged the protest appeared to be "somewhat embarrassed" when they were discovered, although they remained quite convivial. "I have a suspicion that it was a bit of fun and playing around." Advocate Mourant went on to say "If French fishermen had put up the tricolor then that would have been very serious and I would have called the police" however, "My wife and I were annoyed by the sheer effrontery of those who had raised this foreign flag." He added: "The occupation lasted only about half an hour so the Patagonian flag was not in place for long, and his wife had restored the Union Flag to its rightful place while he prepared lunch" All the paraphernalia was removed and handed over to the police' (after Petters, 1998; Tweedie and Jeune, 1998).

The charade of reclaiming the Minquiers as British ... long after the 'invaders' had gone was given prominence in the media with headings ranging from: 'Lone policeman raises the flag to repel invaders' *Daily Telegraph* (Tweedie and Jeune, 1998); 'British bobby retakes Channel isle' *The Times* (Jeune and Macintyre, 1998); 'Silly season brings touch of old Minquiers business' *Jersey Evening Post* (Petters, 1998a); 'Protest: Reef is claimed ...again! Minguiers "farce" gets an encore' *Jersey Evening Post* (Petters, 1998b); 'Jolly good show! Jersey Evening Post (Falle, P. 1998).

The article in the *Daily Telegraph* reported that the invasion was

'too much for the Jersey police. PC Fitchett was dispatched by launch to the Minquiers and used an inflatable dinghy to approach the reef, which has a dozen or so dilapidated granite huts which have served as shelter for ship-wrecked seamen. Taking care to avoid the rotting carcasses of sea birds and pungent piles of bird droppings, he strode up to a concrete jetty. Observed by hundreds of bemused cormorants and gulls, he gathered evidence of the Patagonian invasion, took photographs and watched with pride as the Union flag was hoisted up the solitary flagpole, by David Cadorét, a Parish of Grouville Centenier, and watched by Grouville's Connétable, Frank Amy.'

'The "invasion" force was thought to have consisted of several French yachtsmen, and after raising the Patagonian flag they had nailed tiles painted with the Patagonian emblem to one of the huts, and plastered another building with car stickers bearing the Patagonian colours. Then, judging by the empty bottle left behind, they celebrated their conquest of one of the tiniest bastions of British soil by downing a magnum of Bucks Fizz. They had even defiled the reef's most important architectural relic - an outside lavatory famed for being the most southerly in the country. They placed a notice on the door claiming the historic loo for the kingdom of Araucania and Patagonia' (Tweedie and Jeune, 1998).

PC Graeme Fitchett stated: 'It's too early to see if an offence has been committed, but it's nice to see the Union flag flying ...' 'Hopefully it will stay that way, said Centenier Cadorét'. In Paris, a spokesman for the British Embassy said: 'Honour has been restored.' However, a hiccup in peace negotiations in Paris came when M. Raspail, 73, announced he was prepared to return the filched British flag but only on "neutral territory", by which he meant the bar of one of the most expensive hotels in Paris. "I propose to do it, say on Thursday ...say at the Hotel Crillon or the Hotel Bristol," he said. The Hotel Crillon is where the French Government houses visiting Heads of State, but London baulked at accepting the returned flag in a bar. A breakdown in negotiations was averted, when he finally agreed to bring the flag to the embassy, and so end one of the oddest chapters in modern diplomatic history (Jeune and Macintyre, 1998).

Continuing with the light-hearted spirit in which this whole national flag incident had been treated by officials and media alike, in October, under the newspaper heading 'Jolly good show!' it was reported: 'The actions of the couple, Advocate Peter Mourant and his wife, Anne, who between them "saved" the Minquiers from the Patagonians in the celebrated battle of the offshore reef in August have been recognised officially ... States police chief officer Bob Le Breton decided to recognise the Mourants' public-spirited actions by presenting them with the force's Good Citizen award' (Falle, P. 1998).

It was eleven years after Jean Raspail's second invasion of the Minquiers that on 27 October 2019 The Sunday Telegraph reported under the title 'Invaders claim British rock for a rogue French "king" - A Tiny British Island was invaded in the name of a French "king", with the invaders raising the Patagonian flag and painting a lavatory block in their national colours' (Lowe, 2019).

The paper reported the group approached the Minquiers, a group of rocks nine miles south of Jersey, in an eight-metre boat at "low-light" around 5 p.m. on Wednesday [23 Oct.] in order to carry out their mission. Using a double extension ladder they hoisted the blue, white and green colours of Patagonia, according to Paul Ostroumoff and Julian Mallinson, local hut owners who arrived as the boat was leaving the shore. They were too late to identify them, but saw the group retreat in the direction of France.

It was only later that they realised that something else was amiss. This island's lavatory block – the most "southerly building" in the British Isles – had been repainted in the tricolor of the short-lived "kingdom" established in Argentina in the 19th century. The flag appeared to honour Jean Raspail, a celebrated French author and traveller, who claims the title of "king" by distant lineage to King Orélie-Antonie, the founder of Patagonia.

Mr Mallinson, a Jersey property consultancy director believes that some of his "supporters" were likely behind the action, as Mr Raspail is 94 years old. Mr Mallinson and Mr Ostroumoff, who own two of the island's 10 huts had been visiting the Minquiers to assess the damage done to the buildings after a period of bad weather. Mr Ostroumoff told The *Sunday Telegraph*: "We turned up and noticed that there was a strange flag flying. At first we did not think much of it, but when we got closer we realized it was a Patagonian one instead of the British – so we immediately took it down. Then later I went to clean the toilets and saw the door was painted the same [Patagonian] colour."

The La Maîtresse Île Residents Association regretted that the Patagonian flag had been raised again on the Minquiers, but "Whilst we believe there is no harmful intent by Mr Raspail, we are disappointed that he has painted the hut that is used by visitors to the reef, and has allegedly washed paintbrushes on the rocks, which may disturb the ecology of the area."

Despite of only officially being the king of Arancucania and Patagonia for a matter of weeks, Orélie was credited by some for establishing a parliamentary monarch and an advanced constitution for the indigenous Mapuche people. His legacy has been evolved by French romanticists who have raised the Patagonian flag on the largest of the

Paul Ostroumoff about to take down the blue, white and green Patagonian flag from a post on La Maîtresse Île
Photograph 23 October 2019 ©
Julian Mallinson

Union Flag highlighting Great Britain's sovereignty of Les Minquiers
Photograph summer 2018 ©
Julian Mallinson

Minquiers rocks in defiance of subsequent British claims to sovereignty of the territory. The last attempt came in 1998 when the British flag was replaced and "honourably returned to the British Embassy in Paris". In a statement at the time, the office of King Orélie-Antoine 1 said the "invasion" was a response to Britain's unacceptable and prolonged occupation of the Malouines [Falkland] islands".

On 29 October 2019 he Bailiwick Express reported under the heading 'Minquiers invaders speak out "Brexit made us do it". – A group who 'invaded' the Minquiers in the name of a 150-year-old kingdom have spoken out for the first time since their attempted takeover, saying it was a retaliation against Brexit and revealing their intention to transform the iconic toilet into a 'museum'. The so-called 'special forces of the Kingdom of Patagonia' was the second major intervention of its kind in 20 years … before retreating upon the arrival of hut owners Julian Mallinson and Paul Ostroumoff … her Majesty's Receiver General who is responsible for administering land owned by the Crown pledges to keep the Minquiers "under close review" (Potigy, F. 2019).

Footnote: On 2 October 1999, it had been just over a year since Jean Raspail had for the second time raised the Patagonian flag on the Minquiers. And it had been one of life's great coincidences when my wife, Odette (née Guiton), had the occasion of having Jean Raspail placed next to her at a lunch reception in the South of France. The occasion had taken place the day after the evening ceremony of the 'Grand Prix Litteraire de la Ville d'Antibes' held at the Picasso Museum in Antibes, at which Jean Raspail had received one of the events major literary awards.

Odette took the opportunity to inform the self-styled King Orélie-Antonie that her late father, Pierre Guiton, had owned two ruined properties on the Minquiers, which his family now planned to develop. Also, how she had been well aware of his encroachment of the reefs national sovereignty by his raising the Patagonian flag on such an outpost of the British Isles. However, whilst conversing in French, Odette found Jean Raspail to be an amusing and most agreeable company, who had told her that he had enjoyed placing the British nation on an alert, as well as having had the opportunity to promote his views of Great Britain's occupancy of the Falkland Islands (Mallinson, Jeremy, 2020).

Chapter 7

Guiton Family's barraque

Ownership

La Maîtresse Île has always been a resort of Jersey fishermen. In June 1903, the *'Comité des Havres et Chausées'*, Committee of the States which had been constituted to report on buildings and their owners on La Maîtresse Île, visited the islet. The report recorded: 'They found approximately eighteen houses, the majority of which were built in stone, and inhabited by Jersey fishermen (*barraques*), in which they slept during the fishing season, and one of which was in ruins. How these houses first came to be built, and by whom, it is impossible in most cases to say with certainty. It is however recorded on paragraph 166 of the 'Memorial of the United Kingdom (Vol. 1, p. 95 of The Minquiers and Ecréhos Case, 1953) that during the eighteenth century, round about 1792, workmen in considerable numbers were sent to Maîtresse Île from Jersey to quarry stone for the erection of Fort Regent, and that the quarrying continued during the early years of the nineteenth century' (after Nicolle, in litt. 2005).

In times past, transfers of huts on the Minquiers were not conducted in customary fashion, namely by the passing of contracts before the Royal Court, but rather by private agreement. As Richard Falle relates:

'Rights first to possession would have ripened to property over time as the huts would have passed by operation of law from father to son. Mutations of title by gift or sale were, for long, effected orally with little or no written record. It seems clear however that towards the end of the 19th Century it was generally recognised by the Jersey authorities that the huts and other enclosures were the subject of proper landed title and more and more mutations were recorded by contrat héréditaire passed in traditional form before the Royal Court. The huts are described in the usual way by their tenans et aboutissants - their limits and boundaries - towards the neighbours. They are said to be situated (for want of better intelligence) "sur le fief du Roi/de la Reine ou autre Fief" in the Parish of Grouville on "La Maîtresse Île des Minquiers". Informal unregistered private agreements purporting to transfer title continued to be used into the 20th Century but these became the exception. Examples of formal contracts as evidence of Jersey based title and the Parish rate returns made by hut owners as evidence of municipal administration were filed in support of the English Crown's claim to the Minquiers at the International Court of Justice in 1953' (Falle, pers. comm.). (also see Chapter five).

'Other documents filed by the Jersey authorities include reports of visits to the Reefs in 1888. Some twenty named owners of Minquiers' properties are listed and there is some description of the way of life of fishermen who rowed and sailed from La Rocque every week during the Spring and Summer, staying in their huts and only returning at the weekend with their catches of Lobster. It was this pattern of life which justified the statement 'All who sleep on Maîtresse Île on Census Night are included in the Population of Jersey'. The Minquiers form part of the Parish of Grouville and are accordingly governed by the laws and customs relating to real property of the Bailiwick of Jersey' (Falle, pers. comm.).

In 1996 Mrs Eugenie Guiton, née Besnard, widow of Mr Pierre Guiton commissioned a report from Advocate Richard Falle on her title to two ruined huts which lay immediately to the north of the building owned by the States of Jersey. It was to support Mrs Guiton's application to the Planning Authority. Falle noted that the Guiton huts: 'are those marked 'M' and 'N' ... on a survey prepared in July 1973, by then 'Nature Conservation Advisory Body' for the IDC based upon Rybot's sketch of Maîtresse Île of 1928. The survey noted the condition of the two huts: 'M' – stone walls, west side partially collapsed, gable wall standing to the east; 'N' – stone walls, very derelict, north gable standing with brick stack. Remains of pantile roof evident'.

'Mrs Guiton inherited the huts under the Will of her husband Pierre who after the Second World War, circa 1948, had acquired them by informal purchase from Lieutenant Commander Le Breuilly. An annex to the UK Memorial at The Hague (A98 No. 6) states: "South of (5) and separated from it by a foot-path are two isolated roofless buildings owned by Mr P. Guiton of Gorey, Jersey."

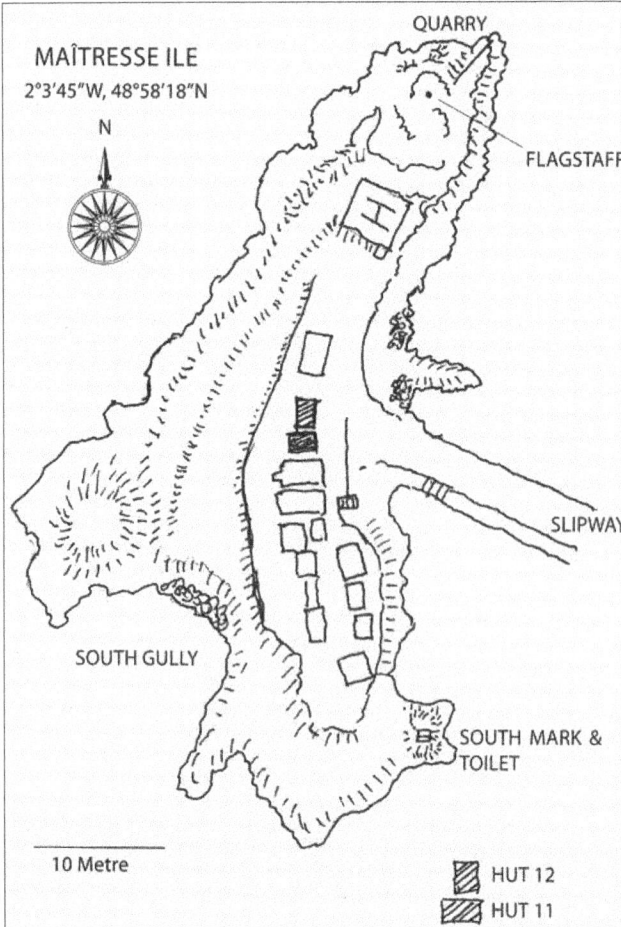

MAÎTRESSE ILE
2°3'45"W, 48°58'18"N

QUARRY

FLAGSTAFF

SLIPWAY

SOUTH GULLY

SOUTH MARK & TOILET

10 Metre

HUT 12
HUT 11

La Maîtresse Île, based on N. Rybot's sketch of 1928. Nature conservation Advisory Survey Body of 26 / 27 July 1973

As to the solidity of the Guiton title, Falle notes:

'The Guiton family is recorded in the Grouville Parish records as having paid rates for this property since the middle 1960's. In the circumstances, I would be surprised if there were any question to be raised concerning Mrs Guiton's title whether...by the States ...or the Planning Authorities.'

*The Guiton family's two ruins on La Maîtresse Île, Les Minquiers
(marked 'M' and 'N' on Rybot's 1928 sketch)*
Photograph 2005 © Andy Hibbs

*Julian Mallinson and Tony Guiton (centre) with stonemason Peter Tyrell
on right, at north end of the Guiton family's ruin ('N') on La Maîtresse Île*
Photograph 2005 © Andy Hibbs

The inference from Advocate Falle's assurance:

'is that any informality resulting from the 1948 acquisition would long since have been cured by the passage of time which establishes a good prescriptive title for anyone who has peacefully and without interruption enjoyed the possession of land for forty years, that is to say "possession quadragénaire'.

The concluding remarks of Advocate Falle's August,1996 letter to the architect, who was about to file an application for development of the Guiton site, highlighted that:

'The Island Planning Law is concerned with issues of planning and development not with private title. While therefore an application to the Committee must be made bona fide by the owner, there is no onus upon the applicant to do other than to claim title, and Falle emphasisd: there is no duty or indeed power on the part of the Planning Authority to demand more than prima facie evidence of such title or to delay consideration of any application against the possibility of a challenge to title being made....In my experience the Planning Officer has not previously raised any matter of title when considering applications for development on the Reef. If however, the matter becomes material I should be happy to support any of the observations that I have made' (Falle, in litt.).

Proposed Development

Sketch proposals for replacing the remains of two huts 'M' and 'N' on Maîtresse Île, and a request for early consent for their demolition, in order to construct a 'Fishing Retreat', on behalf of the Guiton Family, was submitted on 26 August 1996 to the Planning Officer of the States of Jersey Environment & Public Service Committee. The following month the Planning Committee considered the request and wrote that they were prepared to 'see' a single building, subject to a clear architectural plan showing a single storey, with its height, size and scale to be as the building which existed, and materials to be similar to those used previously.

On 27 May 1997, Miss S. Karch, Assistant Planner, States of Jersey

Planning & Building Services, visited the site and, on 16 June, a brief visit was made to the Minquiers by members of the Planning and Environment Committee (P & E). At that time the committee had made a decision to designate both the Minquiers and the Écréhous as Sites of Special Interest. In the latter part of September 1997 Mr Stuart Fell, Conservation Architect/Urban Designer of P & E, and Miss Karch, again visited the reef and as a result of the visit the Assistant Planner advised that they would prepare sketch plans for design guidance. These sketch plans were submitted to Tony Guiton on 30 April 1998.

On 8 February 2001, an application for the development of huts 'N' & 'M' as one unit was submitted to P & E Committee for the replacement of the remains of the two cottages, which was referred to as, a 'fishing retreat'. Almost immediately after this submission several letters of representation were received during February/March by the P & E Committee, objecting to the application. However, despite the fact that the plans subject to the application had been prepared in full consultation with the Planning Department, and no doubt in part due to the objections received by P & E to the development envisaged by the Guiton family, in September 2002, John Tanquy was advised that the application was not acceptable. However, a revised scheme was likely to be acceptable if attention was paid to seven points raised in a letter from the Planning Department dated 25 September 2002. The letter recorded that the Design and Conservation Officer considered that the scheme, as currently submitted, to be unacceptable due mainly to the proposed removal of the existing buildings, and also the scale of the proposed replacement building. In summary, the officer was of the opinion that there were issues relating to both the historic nature of the site, and the ecological sensitivity of the site, that at present would lead to the refusal of the scheme (Farman, in litt. 2002).

Following another visit to the Minquiers by the Planning Officers, and a lengthy meeting on 29 October 2002 with the officers and John Tanguay, a basic agreement was reached on the principles of the application, although certain minor revisions were still required. On 25 November 2002, J Design Limited submitted revised drawings which were based on the profiles prepared in 1998 specifically. Twelve months later, in November 2003, there had been little progress with the application. John Tanquy chased the application, and he was advised that the Committee wished to visit the site 'yet again' in the early spring of 2004 (after Le Gresley, in litt. 2004).

A letter from Le Gresley (2004) to members of the P & E Committee highlighted that despite the fact initial negotiations had begun nearly eight years ago, no decision had been reached by the committee and the Guiton family had asked him to consider how the matter of this application could be brought to a positive conclusion. He also pointed out that when the original application was submitted in early 2001 his clients had been led to believe that it met with basic planning approval. When the revised application was submitted on 8 February 2002, and objections to the development were received, with and regard to the new States of Jersey Island Plan, Mr Gresley's submission was that as the application for development was submitted before the new plan came into being, the provisions of the previous plan should apply. A five - page summary of the negotiations from 1996 to 2004, prepared by John Tanquy on behalf of J Design Limited, was submitted with the letter.

Strip of Land

On 14 July 2004 Mr Le Gresley was advised by P & E to write to HM Receiver General, Mr P.R. Lewin, with regards to the ownership of the strip of land between the two Guiton-owned ruins. Mr Lewin's reply on 11 August 2004 contended: 'that the strip of land between huts 'M' and 'N' belonged to the Crown.' Resulting from this opinion Tony Guiton sought legal advice from Mr Le Gresley, whose 'Opinion' put forward the case that the Guiton family had two claims in respect of the small area of land between its huts: 'The first would be by way of presumption that when the huts were first constructed many years ago the land left between them was land to provide a 'relief' for each of the neighbouring gables. In addition ... the family and the family predecessors having enjoyed both huts for well in excess of forty years without interruption a claim of possession: quadraginaire could be sustainable in relation to the strip of land in its entirety...' Also, a case was cited of two huts 'C' & 'D' on Les Minquiers, belonging to a Frank Lawrence, being joined together in or about 1981; although no reference was made to any land between the two huts in the sale agreement of the 1976 contract between Mr Lawrence and a Mr Philip John Le Claire (after Le Gresley, in litt. 2004; Lewin, in litt, 2004).

In January 2005, HM Receiver General wrote to Le Gresley with regard to his 'Opinion' as to the ownership of the strip of land between the two ruined huts, and with the Crown's interest in the offshore reefs, passed the data to the Solicitor General for further consideration. The response from the Solicitor General highlighted the Crown's 'Opinion' that the huts in question were erected upon land which the builder did not own. Also, that although the possession of both huts for a period of 40 years can give prescriptive title to an area of land, it was the Crown's 'Opinion' that there was nothing whatsoever to suggest that the strip of land between to the two huts form part of the huts.

On 14 June 2005, The States of Jersey Planning and Environment Department issued a Planning Permit to: 'Replace remains of two cottages with one fishing retreat. To be carried out at: Hut 'M'& 'N', La Maîtresse Île, Les Minquiers, Grouville'. However, the only remaining difficulty prior to the development getting under way related to the ownership of the open area of land between the two existing huts. For one of the conditions of the Planning Permit stated: 'The development hereby approved shall not commence until confirmation in writing has been received by the Environment and Public Services Committee that the Receiver General has given approval to the proposals.'

Due to this mandatory planning requirement, and in order to acquire the planning permit at the earliest opportunity, the Guiton family agreed that the strip of land between huts 'M' & 'N' did not come under their ownership. In May 2005, as a result of this understanding, and through further communications between the Crown Officer and N. Le Gresley, 'P & E' approved the development application. And the strip of land between the two ruins was leased by the Crown to the owners of the Guiton's *barraque* for a peppercorn rent on a 99-year lease.

Design & Construction

Mr John Tanquy, a keen fisherman in his own right, produced the plans for a six-person fishing retreat or *barraque* of the site and subsequent development.The overall concept of the design was to produce an aesthetically pleasing and traditional building that would complement the other granite structures on La Maîtresse Île. The granite from the remains of the two huts was to be utilised in the new construction.

The construction team comprised Gary Hollick of G. Hollick

(Builders) Limited, who took on the job of Project Manager, coordinator and general factotum; Peter Tyrell, a well-known Jersey stonemason who was responsible for all the granite work; and Andy Davidson of A & A Carpenters who undertook all the joinery. Apart from being extremely competent in their respective professions, each of the team had years of boating experience which represented another prime requisite for those working on La Maîtresse Île.

In 2006, during the period January to April, preliminary work was undertaken with the Jersey Building Merchant: 'Pentagon', to fabricate the wooden-framed structure of the barraque. The staff at 'Pentagon' proved to be most helpful and generous with their time and can now be proud of having supplied and helped to erect the most southerly timber-framed building in the British Isles!

The question of the team's accommodation was resolved thanks to the States of Jersey giving permission for the use of the next-door 'States' hut, as well as allowing the use of the neighbouring 'Impôt' hut and adjoining yard, for cooking, bedding-down and storage. As it was necessary to have a fast sea-worthy vessel to transport the team with enough food and drink for several days stay at a time, and carry as much building supplies as possible a 25 ft RIB – Indigo Bay was purchased for this task. In order that the team not marooned on the reef they embedded a foundation for a safe mooring to the east of the slipway on La Maîtresse Île, and South Pier Shipyard supplied the necessary chain and buoy. By the end of April preliminary site work was able to begin.

During May each piece of granite from the two ruins was carefully graded and sorted for the intended construction. Foundations were then dug along the lines of the previous walls of the ruins, and when these were completed the Archaeological Section of Société Jersiaise were invited to carry out a 'dig' before the rest of the site excavation was undertaken (see chapter 2 & Clarke, 2010). In compliance with the Planning Permit, Peter Tyrell completed a free-standing section of a granite wall which had to be approved by the Planning Department before further building work could be undertaken. In June Gary Hollick negotiated a delivery of heavy building materials from Jersey on board the States Tug the Duke of Normandy, under the command of its Master, Ian Lamy. The tug was used twice during the course of the year, and without the generous assistance and cooperation from all those involved at the Jersey Harbour Office, it would have been almost impossible to undertake the building of the barraque. By the end of the month the foundation had been completed.

Throughout the construction of the barraque modern day building byelaws were applied by the Planning Department as they would have been for any building work undertaken in Jersey. On 12 June 2006 a visit to La Maîtresse Île took place by the Parish of Grouville Municipality, led by Grouville's Connétable, D.J. Murphy. In a subsequent letter to Tony Guiton the Connétable wrote: 'We were pleased to meet with your builders and have noted the excellent progress they are making' (Murphy, pers. comm.).

In the Spring of 2007 roof gutters and windows were fitted, granite work carefully mortar pointed, and the electrics, plumbing and general fitting-out work completed. During the overall construction of the barraque, Peter Tyrell and Andy Davidson were occasionally joined by other tradesmen to undertake jobs such as lead work; in particular, in 2006 Lawrence Mackie spent several months on La Maîtresse Île, as did Glyn Truman. Every journey to the reef meant the team having to offload heavy material from the RIB to the pier-head and up the slipway and even at week-ends Julian Mallinson and Lawrence Mackie travelled from Jersey with a heavy boat, sometimes in less than ideal conditions.

On the building's completion in the Summer of 2007, its granite work, pantile roofing and overall design presented a traditional appearance that already blended well with the other huts on La Maîtresse. Internally, as intended, it resembled a 'dry-land' boat and has been designed to maximise the space by the use of two sleeping decks within the pitch of the roof, assessed by sliding ladders. An open-plan ground floor includes a kitchenette and a seating area around a multi-fuel fireplace, a shower cubicle, a store and electric plant room. Fresh water is transported from Jersey and kept in an under-floor tank, whilst batteries provide electricity for lighting and the necessary power for showering and hot water, the first time such modern amenities have existed on the isle.

The east side of La Maîtresse Île
Photograph August 2007 © Jeremy Mallinson

Cottages on La Maîtresse Île, showing Guiton family's barraque second on right
Photograph August 2007 © Jeremy Mallinson

The Guiton Famiy's barraque
Photograph July 2019 © Julian Mallinson

Chapter 8

Maîtresse Île
Residents Association

At a Minquiers hut owner's meeting that took place at the St Helier Yacht Club on 10 March 2015, a third draft of the rules of the "Association" were recorded. These included.

Objectives:
(a) To maintain and to preserve the peaceful atmosphere, beauty, tranquility of the Minquiers.
(b) To preserve the natural environment and wildlife of the Minquiers.
(c) To ensure the sustainable use of the Minquiers as a safe and valued environment for the enjoyment of all those who visit.
(d) In liaison with the States of Jersey and other relevant authorities, to ensure that all international and local obligations, especially those associated with its RAMSAR status, are fulfilled.
(e) To assist the monitoring of the harbour area.
(f) To express opinion or make representation on matters relating to the Minquiers.

Membership:
(a) A person who owns or is the spouse or child, over 18, of a person who owns a hut on the Minquiers.
(b) A corporate owner, including the States and the Parish of Grouville, represented by a designated person.
(c) A person who has or is the spouse of, or a child over 18, of a person who has long established use of a hut on the Minquiers, and who is

put forward for membership by an existing member and accepted by resolution ... ...

(d) A Voting Member may appoint a Proxy by written notice to the Honorary Secretary.

Officers and Committees:

(a) To consist of a Chairman and an Honorary Secretary.

(b) The officers and three other Members shall form a committee (the "Committee").

(c) The members of the committee shall be elected by the Voting Members at an annual general meeting, or at a special general meeting, and shall be elected for a term of three years.

As the meeting recognizes that the Minquiers plateau is a unique marine environment, and as an increasing number of visitors are threatening this environment, in order to minimize the impact the "Association" established a Draft "Code of Practice" which all residents and visitors are requested to adhere:

Flora and Fauna – The Minquiers are a RAMSAR site and respect should be shown at all times to the local flora and fauna. Do not feed, touch, disturb or destroy any wildlife. In particular, there must be no disturbance to nesting birds especially during the main nesting season from 1 January to 31 July.

Seals occasionally visit the reef and dolphins are often seen in the surrounding waters. As with all wildlife approach should be made quietly and slowly, not getting too close, with any passengers being told to be quiet as possible to avoid disturbance.

Safety – Tidal currents around the whole reef are very strong. Care should be taken at all times when swimming which, at Maîtresse Île, should be within the anchorage area to the east of the main island, clear of any moorings and advisedly below half tide and above half tide. And great care needs to be taken when landing from and re-boarding any boat or dinghy.

Please be aware of the danger of the rising tide and be sure you do not get cut off on an isolated rock.

Litter, Picnics and Barbecues – All personal litter must be taken away from the reef, including disposable barbecues, whilst disturbance to the general peace and quiet must be kept to the minimum.

Toilet Facilities – There is only one toilet facility located on the main island. Please help us to keep it as clean as possible and only us paper provided. If the toilet is unavailable and needs must, please choose a discreet location below the half tide mark to relieve yourself.

Residents' Privacy – Huts are all privately owned. Please respect the privacy of those in residence.

Moorings – With the exception of the one Visitors' Mooring, all other moorings are privately owned, and visitors should avoid using them as far as possible. If you do need to use a mooring you should move off it as quickly and courteously as soon as the owner arrives. Moorings should not be used by more than one boat at a time except with the owner's permission.

Speed – Speed restrictions are within the anchorage area set out in Jersey Harbours' General Direction No 2 and the 5 - knot speed limit applies. Mariners are asked to show consideration when passing other boats within the mooring area to avoid making unnecessary wash.

Fishing - When fishing at the Minquiers please adhere to Jersey Fishing regulations noting that the General Directions prohibit the setting of fishing gear and trawling/dredging within the area of the main anchorage.

On 14th April 2015 an augural meeting of the 'Maîtresse Île Residents Association' took place at the St. Helier Yacht Club, at which both the draft 'Rules' and 'Code of Practice' of the Association were endorsed. John Le Gresley was elected as the Associations's Chairman, Christiane Gill as Honorary Secretary, with John de Ville, Philip Falle, Julian Mallinson and Heather Speller, as members of the residents committee.

Chapter 9

Personal Landscapes
'Reflections on a Marine Venus'

During the compiling of this publication, in which I have relied so much upon the writings of others, I decided that it would be appropriate and of interest to the reader to present various sentiments and observations from hut owners on La Maîtresse Île, all of whom have been frequent visitors to the reef. And, in particular, to highlight what they consider the special quality of Les Minquiers to be, and the importance of the biodiversity of this unique offshore environment and the need for its conservation.

Martyn Chambers

I first visited the Minquiers in 1951 with my father [Ivor Chambers] in a boat called the Lady Alanda to fish for Ormers. The boat was a very fast ex-patrol boat owned by a man called Sawyer. It was springtime, in March, and it was also my father's first visit to the Minquiers. As it was soon after the War [when no fishing had taken place] there were plenty of Ormers to be had, and after landing on Maîtresse Île, everyone fanned-out and soon returned with sacks of ormers. I don't think that I went back to the Minquiers until after my return to live in Jersey in 1974, when I had my own boat and a group of us used to go out for a week's fishing, usually in September for the high tides. I used to ferry the group out who became the six of us who now share the ownership of our hut [Hut I] i.e. Richard Falle, John Le Gresley, Mike Clapham, John de Veulle, Bernard Morris and me.

The opportunity to buy the hut came about because I had on my staff one of the Le Clercq family. The old man Le Clercq, who was a La Rocque Fisherman, had had a heart attack and had been hospitalised. I believe he was told that he would be unable to use his boat and never fish again and he became very depressed. I took him a catalogue of Cygnus boats which are built in Cornwall, and a set of plans. I had no other thoughts other than hoping the plans may cheer him up. They did, for he soon got his teeth into them and decided to sell his bigger boat and buy a hull and to fit it out regardless. However, in order to do this he had to raise some more money and he decided to sell his site on Les Minquiers which he had owned for a long time. The end result was that I was asked whether I would like to bid for the hut. I think the purchase price was £2,500, which was way above my means at that time. Therefore, I phoned Richard Falle first, and within about one or two hours there were six of us sitting round a table and we decided come 'hell or high water' that we were going to buy it. Although it wasn't quite that simple as one or two other people were quite keen to buy it too. Fortunately, the six of us were able to outbid them.

With the help of the late George Marie who was the former Master of the States Tug, we managed to acquire the adjoining site and to build the hut that you see there to-day. Both huts were derelict, a condition that they had been from before the war, having no roofs and the walls had fallen in. Members of the Boyden family, and one of the Battrick boys, built the huts over a period of about three years about 25 years ago. We would ship them out for up to three weeks at a time, and using the States tug we ferried water, chippings, cement and everything else required. The hut has four bunks and a loft in which there are two more beds. Usually we have been able to borrow another hut so at least the snorers can be separated from those who seek a more peaceful life! Our hut is basic; for lighting we have two pressure lamps, one hurricane lamp

The natural history of the Minquiers is a major attraction; if you go early in the year the shags are nesting and as increasingly are Black-backed Gulls. Other itinerant land birds can be seen as they pass through the reefs. The Mallow there protects vulnerable birdlife as little Firecrests, Goldcrests, Wheatears, various forms of the warbler family, and I saw a Wryneck last year [2009]. With regards to the Mallow, I think there's an urge among those who visit the island for the first time to tidy the whole place up. And certainly many years ago the Harbours Office had a go at doing this with really pretty disastrous consequences. The Mallow conceals a great deal of old rubble and bits and pieces, but it does provide an important shelter for little birds that are making the transit from the continent to Jersey and vice versa.

The force of the sea around the Minquiers in a gale is enormous. In March this year [2010] a very large stone weighing several tons appeared in the centre of the slipway; it had been shifted round the island from one side to the other, from south to the east. I think that well illustrates what we never see, and I hope we never do experience a very angry sea on a big tide. I think it would be quite frightening, if you have a boat at anchor there as well!

The Minquiers represent a wonderful environment. It is wild and temperamental, which is exciting because there are days when we sit with a glass of wine in the evening and watch the sun go down, for the sunsets are terrific, and wonder why we don't come more often. And five days later looking over a boiling sea we wonder why the hell we are there in the first place, The sheer size of the reef at low water and how little there is left at high tide, and witnessing this great movement of water taking place twice a day is really quite breathtaking; there is a great sense of isolation. If you enjoy isolation there is always a spot where you can sit in the sun, read a book and have a glass of wine or just think.

People often ask what the difference is between the Écréhous and the Minquiers; I would say that the Écréhous appear friendlier all the time, for you have the shadows of the cliffs of France and Jersey in sight. In the Minquiers you can just about see the French coast and Chausey, and sometimes there is a feeling of total isolation and wonder in recognising that there is nothing between us and America.

I would like to see La Maîtresse Île left as it is. The islet has evolved over ten centuries, including a giant quarry in the 19th Century and much of the islet is as it was when the quarrymen abandoned it, leaving a lot of rocks and materials lying about. I want to see it protected for the wildlife that use it – particularly birds in the breeding season and during migration.

(Chambers, pers. comm, 2010).

Gordon Coom

During the German Occupation of Jersey my father [William (Bill) Coom] used to go to the Alliance Club in St Helier and on one occasion he won playing cards with the owner of a hut on the Minquiers. As the person concerned couldn't pay what he owed to my father he gave him the Minquiers property, so that's how the hut came into our family's possession. It was originally known as the 'Hospital' [L'Hôpital], because when they did the quarry work at the end of the 18th century it was used as a hospital. The hut consists of two rooms with separate exterior doors which are now both Museums full of artifacts that I have collected over the years!

Just after the war, and within a few days of the Liberation, George Hairon went to the Minquiers on his boat and found three hungry German soldiers who had been billeted on the reef and who had not been informed that the war was over. They had been forgotten! They had evidently survived by eating gull eggs. George Hairon brought the three soldiers back to Jersey where they officially surrendered. During the war the Germans had built a wooden hut adjoining our property, which the States took over and kept until it started to fall apart so they had it demolished.

Soon after the German Occupation my father, the proprietor of the South Pier Shipyard, made the hut over to me. I used to take him in my boat to the Minquiers, which he enjoyed very much. I would go there more frequently to keep the hut in good order and we got Richard Haycock to put a new roof on it. When Fred Le Feuvre, the publican of the Seymour Inn, retired in 1979, he gave me many of his own possessions from the Inn which included the dart board from the bar and I was also given a piano, which my wife used to enjoy playing - the most southerly located piano in the British Isles! Also, I was given from La Rocque Chapel the lectern and one of the church benches (pews) and, since that time, all kinds of people have presented me with bits and pieces and I'm still receiving things of interest for the hut which is now often referred to as 'The Museum'. When visiting the reefs, we used to

a lot of fishing and always slept on the boat, but if we had too many people some of them slept in the hut. My son, William, is keen on the hut and the Minquiers and, although I have still got my boat, my son is now in charge.

Over the years I have taken photographs of some of the well-known people I have taken out in one of my boats to these Jersey offshore reefs. In August 1952, when Rock Hudson and Yvonne de Carlo were filming Toilers of the Sea, I took Yvonne de Carlo to the Écréhous, which she enjoyed very much. One of the scenes from the Bergerac television series was filmed on the Minquiers, and John Nettles and Terry Alexandra with other members of the cast and crew also accompanied me. Marin Marie, who was the first man to sail single-handed across the Atlantic, and the first one to sail back, loved the Minquiers although he was really a Chausey man. He signed three prints of his fine seascape paintings personally for me. During the 1970s I joined St Helier's Jersey lifeboat and served as an RNLI crew member for a period of twenty-five years.

On major Royal occasions, three of us used to place beacons on the reef and Mike Howeson would be in charge of these by setting fire to cooking oil in 50-gallon drums. At the same time, Tim Lough would send a message to the Queen; as an example on Jubilee Year:

> "Congratulations and thanks for the fifty golden years as Queen and our Duke of Normandy who we are celebrating by lighting a beacon on the Minquiers, the most southerly part of your British Isles, and we hope you enjoy your Jubilee year as much as we intend to do."

In May 2002, a letter was received from Buckingham Palace which read:

> "The Queen has asked that her warm thanks be conveyed to you for your kind message of good wishes on the occasion of her Golden Jubilee and received your greetings and congratulations with much pleasure - Sir Robert Janvrin, Private Secretary to the Queen."

One of my favourite recollections was when the Bristol coaster the Brockley Combe (see p. 60) was wrecked on the Minquiers in December 1953 and I went with Bob Viney, Dan Glendewar and Alec Searle on the Lucky Bird, and when we reached the wreck we found about fourteen cases of Bristol cream sherry. Alec immediately mentioned that we would have to be careful for the Customs were obviously going to be interested in the cargo, but we shifted all the Bristol cream sherry from one side of the hold to hide them under a mass of flour bags. The next day customs officers arrived to look at the ship's inventory and asked: 'where was all the Bristol cream?' Alec said that it had in all probability floated out of the hold and sank. While the wreck was still marooned amidst a group of rocks on the reef we would, on occasions, especially at Christmas time, bring a certain amount of the sherry back to Jersey under the floorboards of Lucky Bird. However, as soon as we considered that the Customs were starting to become suspicious, we put some cases on a rock but when we returned to the reef about a week later in bad weather, the cases were no where to be found. We reckoned that as some French fishermen had been previously watching us that they could have well taken them. Bob Viney was of the opinion that they had buried one case on the La Maîtresse Île which has yet to be found!

As the small toilet cubicle on La Maîtresse Île is the southern-most building in the British Isles I considered it appropriate to have a notice made to put on its door to read:

THIS TOILET HAS THE DISTINCTION OF BEING THE MOST SOUTHERN BUILDING IN THE BRITISH ILES PLEASE USE IT WITH CARE! AS THE NEAREST ALTERNATIVE IS JERSEY 11 MILES CHAUSEY 10 MILES

(Coom, pers. comm. 2010)

*

After the death of Gordon Coom in 2017, his property was inherited by his son William. In the spring of 2018 Advocate James Lawrence and senior conveyancer Michael Falle, of Viberts, visited the Minquiers for the sale of L'Hôpital, which was sold to two Jersey-born residents (JEP, 2015 – see Ostroumoff & Painter.

Richard Falle

I graduated to Jersey's offshore reefs from low water fishing at La Rocque as a child. Seymour Tower had been an icon for me and my friends in the 1950s and 1960s. We used to stay there for a tide in August or September - pêche à pied during the day, dragnetting at night. After Seymour we fished the Écréhous for a few years but found that Reef somewhat crowded. The Minquiers in contrast, enticingly stretched out on the southern horizon, seemed to us to have a more solitary and challenging character. Getting there was altogether more difficult.

I was in my late twenties before I had my first taste of the Minquiers on a day trip with John de Veulle courtesy of a fisherman's boat. I remember we were very successful that day coming away with a catch of about 60 Ormers each – unthinkable half a century later.

I first stayed overnight with John Le Gresley at the Minquiers. We were the physically active "youngsters" attached to a party of elderly men who showed us where to drop pots, where to prawn and where to line, although the location of secret houles remained a jealously guarded secret. These men - friends and relatives - included Jack Le Gresley, Bertram Payn, Jim Le Montais and others. From them we learned the old Jerriais names of the rocks and shoals.

After those first visits we, John Le Gresley and the original Seymour Tower group would stay at one or other of the two stone huts belonging to the States of Jersey. A key was always readily available on request by weekend fishermen such as us. The older hut, the Impôts, graced with the Arms of the Bailiwick cut in stone dates from the end of the 19th Century; the other, a more imposing stone structure administered by the then "Harbours and Airport Committee" was built in Jersey granite in contemplation of the case on sovereignty before the International Court of Justice in 1953. These official buildings clearly had no other function than to annoy the French!

In about 1980 we were very lucky to acquire the sites of two small adjoining ruins and a couple of years later, with the lovely weathered stone from the broken walls and using old pantiles, we constructed our present hut on this footprint. It takes the form of an original fisherman's hut, writ large.

The issue of land ownership at the Minquiers is not entirely resolved. In particular, HM Receiver General has, in recent years, taken a new interest based on a supposed Crown claim to the rocks. The Crown, it appears, would base its right and title on some kind of possession from "time immemorial". I personally have undertaken research on the matter and believe that claim to be ill-founded. It is not in issue that the Minquiers Reef was formerly part of the Fief of Noirmont and, in the Middle Ages, a possession of the Abbot of Mont St Michel. The Fief came into the Crown's hands with the seizure of the Alien Priories by Henry V. Noirmont was later granted to George de Carteret in the 1640's as a reward for his naval exploits in the Mediterranean. There was no reserve in that grant and the Minquiers, valuable for its fish, rights of wreck and as a quarry must therefore have passed to de Carteret with the Fief. Noirmont has since passed down by inheritance, purchase etc. and each mutation of title has been ratified by a formal Order in Council. The Crown cannot derogate from its own grant. The Seigneur is now Leonard Jagger of Noirmont Manor. In my opinion, therefore, Mr Jagger, as Seigneur, has title to the soil of the Reef.

I have good authority for the above statement. In 1953 at the International Court of Justice it was successfully argued on behalf of the United Kingdom Government and Jersey against French claims that the Minquiers were parcel of the Bailiwick because, 'inter alia', they fell within the jurisdiction of the Fief of Noirmont. Proof lay in the Noirmont Rolls which record the Court sitting to adjudicate a number of interesting 16th Century cases of wreck landing on the Minquiers Reef. Territorial jurisdiction was one of the planks on which the ICJ based its Judgment. The Noirmont connection was accordingly an important part of our case but the Crown's claim was limited to sovereignty not ownership of the Reef. It was never part of the Ancient Domaine nor indeed, in my opinion, could there have been any other basis for Crown title.

What then of the hut owners? My friends and I purchased our two sites and passed formal contracts before the Royal Court. We were by no means alone in that. A great deal of evidence of the same kind was before the ICJ to support local titles including Parish rate returns and contracts of purchase and sale. It was the case of the Crown in 1953 that good private titles to land existed.

Staying at the Minquiers for a fishing tide is a wonderful experience. There is a sense of being totally cut off and isolated from the outside world. One is constantly aware of the extraordinary tidal movements and the immense currents surging to and fro in and out of the Bay of Mont St Michel. A very wide sky hangs over a remarkably horizontal world. This is a perfect place for reading, conversation and contemplation.

At half-tide on the ebb there is always a gathering of forces to discuss which part of the Reef to fish and what equipment to take. We usually go out miles into the middle of the Reef, leave our boat in some sandy cove, have a hasty lunch then fan out with prawning nets, lobster hooks and sand eeling rakes. A couple of hours later we rendezvous at the boat returning to the Maîtresse Île on the rising tide.

Sundowners on the western rocks watching the sun go down on Les Maisons at the far edge of the Reef is a hugely enjoyable experience. We often see passing dolphins, the odd seal or just watch the birds as they crowd the few uncovered rocks. When the light fails, we return to our hut to consume heaps of hot prawns, lobster, salads etc., washed down, of course, with a plentiful supply of wine.

My friends and I are tremendously fortunate to have secured a ple on the Reef and lucky that the owners of the other huts also respect the spirit of the place. We find the French, whether yachtsmen or fishermen, whom we meet on the Reef, to be almost invariably civil and friendly. Most of us admit to being under the same spell.

(Falle, pers. comm. 2010)

Juliet Gillam

My first visit to the Minquiers was in the early 1960s when my father, Advocate Vivian Vibert, borrowed the States hut for a few days. Our family sailed out in the Fiona, an open gaff-rigged Jersey oyster fishing boat. The hut was rarely used and very basic. It had a big table, one long bench and a little cast-iron cooking range. The island was covered in shoulder-high Tree Mallow and there were seabirds everywhere, wheeling overhead and running about on the ground. It was a wild place untouched by man.

Over a period of year's we visited the Minquiers many times, often with my sister Christine and my brother-in-law Vincent Obbard, who was renovating his hut La Pointe. Then one day Peter Baker, who was a great friend of my father, said to him that if anyone in our family wanted to do up his hut he would be glad for it to be used. My husband Richard and I thought it was a wonderful idea, and very generous of Peter. In

1979 we had it rebuilt very simply and went to the Minquiers frequently, often with Vincent and Christine and our small children, sharing both the huts as they are conveniently adjacent. Peter Baker died shortly after my father, and much to my surprise he left me the hut in his will.

Every year we had an extended stay on the Minquiers, taking the children even when they were very small. I would take the pram and park it at the top of the steps near the slipway with the baby in it. One day some French yachtsmen landed, saw the pram with the baby sitting in it and said "Ah! L'Amiral Nelson!" On one trip we stayed for three and a half weeks, at that time there were very few visitors and we had no contact with the outside world. When staying over an extended period time becomes irrelevant, and one's daily activities are governed by the state of the tide, there is a wonderful sense of freedom.

When Maîtresse Île was largely unvisited, and most of the huts were in ruins birds dominated the island. On landing, particularly in the nesting season, the island reeked of guano which was feet thick in places. Shags nested on untidy piles of mallow sticks in amongst the ruins, and in the gaps between the Impôt and States hut, and the Guiton hut. The Le Masurier hut which was derelict but still just about standing was full of young Shags 'oinking' and squawking - we used to call it 'The Pig Hut'. All around were Herring Gull nests with fluffy chicks, and oystercatchers nesting on the higher ground near the two flagpoles. There were very few Great Black-backed Gulls, which are now the only birds bold enough to continue to nest on the main island.

The lack of nesting birds has had an adverse effect upon the Tree Mallow which is gradually getting smaller due to the lack of guano. It is also a great pity that people uproot the Mallow to 'tidy' or to make room to pitch a tent. The Tree Mallow is essential for small migratory birds, with Chiffchaffs, Firecrests and warblers passing through in the spring and autumn in numbers, and also many other species of migrating birds.

I would like the whole Plateau des Minquiers to be designated a Marine Park and conservation area, where fishing is prohibited, which would allow respite from human interference to all marine creatures and birds. I would be happy to forego all fishing activities in the interests of wildlife and would also support a ban on people visiting Maîtresse Île during the nesting season, thus allowing the bird population a chance to re-establish itself. These views may seem extreme, but strong, positive and unselfish action should be taken to protect this extraordinary habitat.

One thing that I find very special about the Minquiers is that even if we all come and go, rebuild huts, mess about in boats, there is an eternal quality, always the same rocks, the same sea going ceaselessly up or down - it doesn't matter when you go, whatever time of year, whether it is lashing with rain or a sunny day, the essence of the Minquiers is that it is unchanging. I feel very privileged to have been able to experience such a magnificent environment, the visual beauty of the Minquiers is outstanding and I will retain forever strong images of shapes of rocks, the dramatic outlines of the reef extending to westwards, and the ever moving sea.

(Gillam, pers. comm. 2010)

Andy Hibbs

I first visited the Minquiers with my Dad, and Gordon Coom, in the late 1970s. Also, in the early 1980s, we frequently went to the reefs with Brian Maguire and Frank Lawrence. My Dad was a member of the Lifeboat crew in St. Helier with Frank, so when Frank was building his hut on La Maîtresse Île we used to deliver building materials for him. It was great fun aboard my father's boat the *Barberella*. And it was funny that I never thought I would end up fishing there and living in Frank Lawrence's house.

As a youngster I was always amazed how the reef changed from tide to tide, the tranquility, peacefulness, sheer beauty of the place, and yet it can be so violent and dangerous at times. When I started fishing during my holidays with Steve Maguire, it was a brilliant place. I acquired my first hut in July,1996, and purchased the second hut from Chris Le Boutlier in 2015. I always wanted to have this hut, for it faces west with an amazing view of the setting sun, which we intend to rebuild soon.

It is an astonishing place and the older you get it becomes a part of you. And it has been a massive part of my life, and I hope it will be protected for future generations to come. When you have spent so many years there like I have, in all weathers, you really appreciate it. There is no better place in the world on a fine summers' day, at low water.

With regards to having witnessed problems/disasters on the reef I have seen fishing boats sink, boats washed ashore, and some boats arriving thinking that it was Jersey, whilst others in thick fog thinking it was Chausey. We did a lot over the years with the Lifeboat Medical Emergencies, etc.

But the worst experience that will always live with me was when I was stuck there in 1990 for two weeks in horrendous weather, gale after gale. This was at the time when the French Fishing boat *Revolution* sank on 25 January, with the loss of all hands, between Chausey and Granville. I was listening to the tragedy all unfold on the radio.

(Hibbs, in litt. 2020)

Captain Frank Lawrence

Salt water has been very much in my family's blood. My great-grandfather, grandfather and father all went to sea. My father had been harbour-master at Gorey prior to the Second World War, so I was going to sea whatever happened.

It was my grandfather, Francis Henry Lawrence, who first became directly involved with Les Minquiers, for in April 1910 he had gone with a friend to the reef to fish. They had anchored their boat close to the Pipette reef to lay a 'trot' [a fishing line with a lot of hooks]. Later, at dusk, they rowed out in their small dinghy to retrieve the trott with its catch. They had left a light on board the boat, either an oil lamp or a candle, but when they returned to where it had been anchored the boat had disappeared. As my grandfather was overdue back in St. Helier by two days, with no news having been received from the two men for over four days, the crew of the States Tug, the Duke of Normandy, was instructed by the Harbour Master to make a thorough search of Les Minquiers and the neighboring reefs.

However, prior to the search being able to take place ship engineers had to carry our some essential repair work on the tug and they, seeing the priority, gallantry carried out the work in the early hours of the morning in order for the Duke of Normandy' to make the trip to Les

Minquiers at the earliest opportunity. It was shortly after 2 p.m. the following day that the tug returned to St Helier with the two missing men on board, who had been found safe on La Maîtresse Île. The men told a thrilling tale on landing, for when my grandfather found his boat to be missing, and the wind had started to blow-up a little, they rowed through the rocks from the Pipettes to Les Maisons reef. Here they remained for two days, being extremely cold, wet and very thirsty, prior to calmer weather when it was safe enough to row the four miles to La Maîtresse Île. They were certainly fortunate in having escaped with their lives. With the loss of my grandfather's boat the local press highlighted that as it had been his only means of gaining a livelihood for his family, they generously launched a public appeal

in support of a replacement. Various headlines appeared in the Jersey press: 'Missing Fishermen – Tug to the Rescue', 'No food for Two Days' and 'Without Food and Water' (Morning News, 1910; JEP, 1910).

I first started going down to the Minquiers when in 1951 I got a job on the States tug, the Duke of Normandy. We used to carry out the maintenance work on the buoys and beacons around the Island and the offshore reefs. Initially, I worked on the States tug for a period of six years, and I got to love the reefs and I always wanted to buy a place down there. It was during this time, in August 1956, while working on the Duke of Normandy that I was involved in a rather macabre incident in having to pick up the body of an unidentified middle-aged woman off the Minquiers. This was reported in a rather dramatic article in the Jersey press under the title: 'Body Picked-Up off the Minquiers' (JEP, 1956). In 1957, after leaving the States tug I worked on coasters before joining the crew of a number of the British Railway mail boats where, through a correspondence course, I secured a Mate's Certificate prior to rejoining the Duke of Normandy. It was during these early days back on the States tug that I was given the opportunity to study for and to gain a Master's Certificate.

The only slight mishap that I experienced while being Master of the old States tug, when we had no radar or GPS and we just relied on timing, was in August 1966 when we were going down to the Minquiers in thick fog to inspect the States buoy at the Demie de Vascelin reef. I stopped, on time, with the engines no longer running and we were just drifting, but in the fog still unable to see the buoy. The next thing we knew we were on the rocks, which were just under the water's surface. On impact, I put the tug to full astern and it immediately freed itself and I anchored. This incident was reported by the Jersey Evening Post: 'States Tug Hits Minquiers Rocks, Lifeboat Stand-By'. The article recorded a comment from Mr G.S. Seymour senior, a passenger on board: "The Captain acted with commendable calm. The vessel listed suddenly but soon righted itself and we were not sure whether we had taken on any water.' The lifeboat escorted the tug back to St Helier Harbour and the tug was later found to be undamaged." (JEP, 1966).

Some years later I was invited by the Town Pilots to be a partner in the Company where I remained for twenty years. And it was while I was a Pilot that I managed to acquire two ruins on La Maîtresse Île; for in those days the majority of the huts on the reef were in a ruined state. As Les Minquiers come under the jurisdiction of the Parish of Grouville I went to the Parish Hall to make enquiries, and they referred me to Edmond Le Claire at La Rocque who knew my family. When I asked him whether he would be willing to sell his property to me he immediately replied: "Yes, certainly Frank, I have two ruins alongside each other and you can have them both!" As there was a gap of about a 6 ft between the foundations of the two ruins, in order to be able to construct a single building I had to gain permission from Jersey's I.D.C. [Island Development Committee]. Although permission was soon granted it was on condition that the structure kept within the same footprint of the two original foundations.

I bought my boat in 1977 especially for the purpose of building the hut, so that I was able to carry the tons of stone, sand

and cement needed for the hut's construction. Dennis Aubert, with the help of two others, which included my younger son, Francis, who was the apprentice, completed the building for me by the autumn of 1981.

I have always greatly enjoyed working on the Minquiers, for I liked the pilotage and navigational side of things, and almost everything about these isolated offshore reefs. However, I fully recognise how very fortunate I am to own a property on Maîtresse Île, but unlike my grandfather I am not a fisherman, for while at the Minquiers I normally spend my time working on the house and checking my moorings. Members of my family have always enjoyed their stays on the reef and one of my sons, Captain Peter Lawrence, as a Master Pilot and an Assistant Harbour Master, has continued the family's seafaring tradition to the fifth generation. Now, I only wish that I could spend more time there.

(Lawrence, pers. comm. 2010)

Footnote - The Jersey Cruising Guide for 2010, which is given out by the St Helier Yacht Club, has an excellent article by Captain Lawrence entitled: 'Pilotage Notes, the Best Way to the Minquiers'. The article includes a most valuable navigational guide on how to get there, with six different colour diagrams.

Julian Mallinson

I grew up with stories about the Écréhous and the Minquiers, knowing that my grandfather Pierre Guiton had owned huts on both, but not of anyone who had actually visited. In those days the Minquiers were from a bygone era, a time before GPS and VHF, when only the most salty of seadogs could navigate the reef's safe passage. The words 'treacherous' and 'dangerous' became synonymous with the Minquiers.

From our family home in St Aubin, on a clear day and at low tide the Minquiers would expose itself as a rocky outcrop upon the horizon - a mythical island. Unfortunately, my family's nautical past disappeared with my grandfather, and I grew up with a rubber dinghy, windsurfers, but no boat. Therefore, it wasn't until my early thirties that I visited the Minquiers for the first time, on a summer's evening, aboard Tim Le Gallais' RIB. However, whilst having a barbecue next to the two ruins owned by the Guiton family, I knew it certainly wouldn't be my last.

Fast forward ten years, and I was now captain of the Guiton family's 26ft RIB battling against the elements while transporting tons of building material to the reef; rather a steep learning curve and an even a steeper slip to surmount! It has now been thirteen years since the building of our hut has been completed and I have enjoyed many memorable times from brief day-trips to spending a week at a time on La Maîtresse Île, sometimes with only the gulls as companions. When I talk about the Minquiers I am often asked "Don't you get bored?", and "What's there to do out there?" Questions that I still find strange.

Time spent on the Minquiers is always busy and the days are entirely governed by the tides. Each new tide provides windows of opportunity between its ebb and flood, periods of time which if lost will necessitate several hours wait, be it pulling a boat alongside the jetty or foraging for food. At low Springs a plentiful supply of shrimp may be found sheltering beneath the vraic, if you're lucky (or seemingly have the visiting French knowledge) crab, lobster and the elusive ormer may be found in the rocky crevices. Once the tide starts to flood razorfish rise from the sand, and if you tiptoe and cast no shadow these may be grabbed before they shoot back into their holes. These, accompanied by raked clams and garlic butter, join the shrimp on the 'barbie'. For me the first few hours of the flood are the most awaited part of the day, for it's prime time to trick bass into striking at lures masquerading as injured fish. There have been many an occasion when the barbecue has been lit at dusk and a bass caught, gutted, cooked and eaten within twenty minutes… and no fish will ever taste so good!

Paddle boarding is spectacular at the Minquiers and I particularly enjoy circumnavigating La Maîtresse Île at half-tide, gliding over translucent water observe fish darting amongst the marine forests, then battling against the currents to the static wave, sometimes avoiding an over inquisitive seal.

On longer stays, and when the tide isn't too high, the lobster pots need to be inspected and re-baited which depending on sea conditions is a mission in itself. In fine weather during low Spring tides the RIB is taken out to explore the sandbanks, shipwrecks and remote outer reefs. One can anchor in the centre of these reefs in complete solitude without seeing any human activity, truly the last Channel Island wilderness.

It is only at high tide that activities slow down, the reef shrinks to become very small for the first time and one becomes conscience that all that remains is a rock in the middle of the sea. Time to open some bottles and observe the black-backed gulls dogfighting above. As the tide reaches peak, flocks of oystercatchers, terns and colourful migratory species gather in groups on the highest rocks, forming a natural amphitheatre, chattering excitedly, whilst waiting for the human performance to commence ….

For me the reef is from a parallel earth, where clocks become irrelevant and the weather and tides control the day. It evokes the hunter gatherer and self-preservationist within a visitor, its beauty has an edge, its danger is invigorating.

I have always wished to be dropped-off on La Maîtresse Île before a great storm. Past storms have caused waves to surge in-between our hut and L'Hôpital and 5-ton boulders to be swept up the slip. I imagine being barricaded within the hut at high tide surrounded by the thunderous sound of sea and rock and each summer I await a deep depression coinciding with a high Spring tide - the perfect storm!

The Minquiers may on occasions still be "treacherous" and "dangerous" but to me they are neither anymore.

(Mallinson, in lit., pers. comm. 2010 & 2020)

Vincent Obbard

My Minquiers hut is known a s La Pointe and is marked 'B' on Rybot's 1928 sketch. I acquired it in 1973 from Roselle Bolitho (*née* Lemprière), the widow of my Godfather, Captain R.J.B, Bolitho, who had acquired the hut in 1926.

How did I have this good fortune? Captain Bolitho was a close family friend to my parents during the Second World War in London. After my father's death in 1949, he was a frequent visitor to my mother and me at Samarès Manor and we would visit him and Roselle at Rozel Manor. He was a good and probably strong influence in my formative years. He was always keen to instil in me the joys of sailing in local waters and we made frequent visits to his hut on the Écréhous. I loved his sailing stories and his account of his experiences in both World Wars. Well over six feet tall, he would be accompanied by his highly trained dog 'Jacko' (or, later, by 'Toby' both large German Shepherd dogs). At one time, he had owned a 12-metre racing yacht, equipped with crew, but no engine, which he courageously sailed around Channel Island waters. He was credited with inventing a radio aerial suitable for a First World War military tank and for designing naval gunnery range finders in the Second World War. A story I particularly liked was his memory of standing in the crowd outside Buckingham Palace on VE day, shouting "We want the King!" and finding himself standing on a flower bed of geraniums. He picked a bunch of stems which he kept alive in London in a tooth mug and eventually brought back to Jersey with him. He would annually have his own flower beds at Rozel Manor planted with "Buckingham Palace" geraniums. You could say he was a little eccentric, but a more lovable man you could not wish to meet.

I never went with him to his hut on the Minquiers which he regarded as "off limits" for family enjoyment. I believe that he had originally acquired the hut in order to go goose shooting [Brent Geese]. It was not until after his death, in 1965, that I happened to speak with his granddaughter, who said I should speak to Roselle about the Minquiers Hut. To my amazement she said, 'you can have it if you like'. She must have thought that my late Godfather would approve. Roselle was a client of the legal firm where I was working at the time and studying to become an advocate. I lost no time in drafting a deed of gift which was duly placed before the Royal Court.

I first went there with my friend Jim Hardy. It was a terrible trip. We didn't really know the way in. We spent the night sleeping on board in the anchorage feeling sick. The hut was a semi-ruin. The roof had partly fallen in over the added western section and there was major repair work to be done.

Next to the hut is a toilet which I had been informed was built by my godfather. I always knew him by his First World War nickname 'Pull-through' and have always known the toilet as 'Pull-through's loo'. It is painted as a shipping mark and makes a passage from the anchorage due south at half-tide.

The legal firm I referred to above was then known as Vibert and Vibert. At the time I was going out with Christine Vibert, the daughter of one of the partners, Bob Vibert. I was entitled to study leave for my Jersey Bar exams. I regret to say that we rather took advantage of that privilege by going on frequent sailing trips on my boat Peri of Keyhaven, which included trips to the Minquiers taking wood for the door and window frames, slates for the roof and glass for windows. Over the years I fixed the hut as I imagine it would have been in Uncle 'Pull-through's' time.

We were once stormbound there in September 1974. The forecast wasn't too bad when we left Jersey, but it blew up to gale force. We entrusted Peri to my new mooring chain which fortunately held. We were marooned for five or six days. We stayed in the States hut with very little food because La Pointe was not yet habitable. We ate limpets and wild Sea Kale and drank rainwater. We tried fishing without success. We had a ship to shore VHF radio and made 'link calls' to Christine's parents. Eventually, the States Launch, the Duchess of Normandy came out with supplies for us. We were given the option of returning on the launch to Jersey, but, as the storm was by then subsiding, we opted to return on board Peri.

There was this amazing holiday which Christine and I had with her sister, Juliet and Richard Gillam, I guess in 1974, and I think it was at that stage we fell in love with the place. It was before anyone else had refurbished their huts. We caught crabs and we would steam them in seaweed over an open fire in the fire place of a ruin. It was a fairly crude way of cooking but somehow it was right at the time and we ate all this half-raw fish without stomach upset. Juliet was always a good fisherwoman for she used to catch fish when nobody else could. Going out there by boat, the solace of this incredibly remote place which was hardly ever visited by anybody at that stage, and the knowledge that the reef extended at least five miles out to the west and we hadn't really explored any of that, this was all incredibly exciting. It was like we had a foothold on the edge of the earth.

I remember two special trips. One, when Christine and I decided that we were going to take her father's boat Bricoleur right through the Minquiers at low tide. This was quite an adventurous thing to do for we really didn't know our way and it was a neap low tide. We managed to go through the main passage in the centre of the reef with very little water under the keel and exited to the west. Another with Richard Falle and John Le Gresley, when we caught various lobsters and ormers, and a wonderful time when a shoal of mullet came so close into beautiful clear water by the harbour jetty and we were able to catch the mullet only a few feet away.

There is a rock that I know where you can put your pot down and if you catch a lobster you think, 'Hurray! I've got a lobster,' and perhaps even carelessly you decide to throw the pot back and collect it at the next tide. You come back and at least you find one or two more lobsters in the same pot in the same place! Marine life is really abundant. Commercial fisherman from both Jersey and France do fish the reef for shellfish there but they tend to go in larger boats and some of these little inaccessible secluded corners still yield fantastic catches.

I completely re-roofed the hut in 1990 with the help of my then girlfriend, Gillie, who is now my wife. The roof used to have orange coloured asbestos slates which must have been originally done by Captain Bolitho. I wanted to get the original coloured slates but I couldn't find any so the new roof had to have grey ones. Richard Gillam gave us invaluable help.

What of the future? My son Edward has the same enthusiasm for the Minquiers, if not more, and has spent substantial time there living on his own, but is now working and living in Australia, so visits are intermittent. Meanwhile, I think that the Minquiers 'Ramsar' status should be taken seriously and respected by government. Not only is it a very special place. What makes it so, needs protection. The islands are a unique staging post for many unusual species of migrating birds and a hunting and breeding ground for a large range of different gulls. A colony of Seals has made it their home. The fish life is amazing and abundant and even the microclimate of the vegetation on Maîtresse Île is a delicate balance between the scarce surviving earth, the salt wind and spray, the sweet sun and rain and the vegetation and creatures which can survive in those conditions.

Archaeological digs have revealed evidence of how life was at a time the island was not surrounded by sea, but a plain, joining the Minquiers to Chausey and France. More recently, the principle reason for development of the island was stone quarrying. The art of how bargemen negotiated the rocks with their craft and tied up close enough for stone to be loaded by winches and pulleys has fascinated me and has never been fully investigated. Many of the rocks on the Island have been inscribed by quarrymen giving their initial and the date the stone above was removed with iron or formerly wooden wedges. Several of these stones date from the 18th century. Legend has it that the fishermen were so enraged by the gradual destruction of the anchorage and surrounding reefs that they seized the quarrymen's tools and threw them into the sea, but not before much of the north of the island was removed by explosives. For my part I am so lucky I have had the opportunity to enjoy the remoteness and unique position of the reef at more peaceful time. The memory of my uncle 'Pull-through' is never far away.

(Obbard, pers. comm. 2010)

Paul Ostroumoff & James Painter

Paul's first visit to the Minquiers was in the 1980s with his grandparents, although they never went ashore: 'My next visit had been with Julian soon after the Guiton Family barraque had been completed in 2007. On a number of occasions James (Painter) and I stayed in the hut, but we never dreamt that one day we would own a hut on the Minquiers ourselves.

After the death of Gordon Coom in 2017 Julian mentioned to us that Gordon's property, known as the L'Hôpital, could well soon come on the market, and he was a bit concerned who his new neighbours might be. So, after taking some advice from Julian we (James Painter and I) presented a sensible offer to Gordon's son, William, and hand-delivered a letter to him at his base at the South Pier Shipyard. As we did not hear from William for over a year we started to be slightly concerned, for we had heard that other parties were interested, so we decided to get in direct contact with him and we had a meeting with William at the St. Helier Yacht Club. At first, he could still see his Dad 'waiving his hands in the air' and was not too sure whether he wanted to sell, but about a week later he phoned to tell me that the deal was done with us, and that was amazing news!

When we purchased the hut I was asked by a well-known St Aubin character "what on earth did you buy that for ?", a bit taken back I responded " well, we live on a hectic island, I've purchased a pace of life which we all crave for".

I did feel after we had purchased the hut whether we had done the right thing, for soon after acquiring it we had a couple of particularly bad journeys home, but as Julian mentioned, 'If you don't experience such sea conditions, they could well catch you out in the future when you're by yourself'. We have now bought "Seawarrior" a good, seaworthy boat. And soon after its purchase I was a guest with my mother [Diana Martland] on Martin Richardson's boat on a trip to the Minquiers. However, although going around the island in a dinghy, due to sea conditions, we were unable to land and actually lost the dingy (recovered the next day) which was a timely reminder of the currents surrounding the island.

The best experiences are hard to say but to go out there at the beginning of the season and to see seals, and a great variety of bird life ranging from puffins, herons to an owl, is a brilliant time to visit. There's a magic to the reef, which at times is beyond belief, for to me it's the jewel in the crown, if Jersey is the crown within the British Isles. I have spent time in the Caribbean and many other places, and there is nothing quite like it. The quietness of the reef, and its cheer beauty, with the diverse colours of the rocks; its turquoise waters; and at low tide the seemingly 'munching' noise of all the creatures on the seabed; and with the skies at night without the light pollution, are some of the best that I have ever seen. I am looking forward to going out to the reef as soon as the 'Covid-19' virus lock-down is over, for without any planes flying around there is apparently a good possibility of seeing Jupiter.

With regards to our plans to renovate the hut we are keen to keep it as traditional as possible, with Welsh slates on the roof, lime plaster work, oak doors and windows. It would of course be lovely to have the hut with a thatched roof, as it was in the early days, but this is obviously no longer feasible.

*   *   *   *   *

Growing up in Jersey some of James's earliest memories were sheltering on the cockpit floor of his parent's boat as they battled the seas between Jersey and the French Coast: 'With navigation often relying on the siting of cardinal buoys, the word "Minquiers" was frequently heard as we sought out the yellow and black buoys that mark the compass points around the reef. However, in those days before GPS, my parents would never dare venture into the reef (seeing as it is such a hostile place for a sailboat), which can easily be taken off course by the rapid tidal flows in the narrow gulleys.

These regular sailings passed Les Minquiers gave me a great curiosity for the reef, so I was delighted to have the opportunity to finally visit when the Guiton family began renovating their barraque which lay in ruins at the top of the slipway, on MaÎtresse Île.

On my first visit to Les Minquiers by RIB, the biggest surprise to me was how easy it seemed to navigate to Maîtresse Île, despite its treacherous reputation. What I did not realise at the time, is that our captain on the day knew every part of the reef so well that he could have probably have taken us in with his eyes closed, so made it seem very straightforward; however his knowledge and skill belied the truth.

In reality, navigation around the reef is far from straightforward. Many prominent and dangerous rock heads are unmarked on charts and dry out at as much as half tide. Further confusing the uninitiated are the rapid currents creating eddies where no underlying hazards exist, yet calm areas where they do. Consequently I have long believed that a good pair of polarised sunglasses are of more benefit to the art of navigation than a state of the art GPS, as looking into the water will give you far more information than looking at a screen.

One feature of the reef is that at high tide only two notable land masses remain above water. The habitable Maîtresse Île to the east and the barren mound of inhospitable rock to the west named La Maison. Lying some three miles apart, and looking very different at low tide, they have an almost identical appearance at high tide. This similarity caused a group of us a problem on our first attempt to paddle board to the reef some years ago. With the strongest and leading paddler having incorrectly entered the GPS numbers for Les Maison, we did not realise we were paddling to the wrong land mass, until our support boat radioed us to tell us we had gone 3 miles off course on a 12 mile trip. Whilst kayaks are certainly more popular than paddle boards, I have since heard that the journey has been made by paddling surfboards back in the 1960's and I would be thrilled to find out more details if this is true.

Having been fortunate over the past 10 years to make many visits to the reef I was delighted when Paul Ostroumoff and I had the opportunity to take over the stewardship of the barraque known as L'Hopital from William Coom. Under the Coom family the hut has earned a rich history and an elective collection of keepsakes. Our main aim, when renovating and making the hut habitable again, will be to preserve the Coom legacy and add to the heritage of this historic building, in what is truly a unique location.

There are many reasons why I feel such a bond with the reef, the crystal clear water, the natural beauty, the excellent fishing, and the delight of having no phone or internet signals are all great pulls, but it is the connection with the tide that I love the most. Whenever visiting Maîtresse Île, life is ruled by the times, direction, and size of the tide, which dictate what you do, when you do it, and often what you eat.

So if I had to choose one reason above all others why I love to stay at Les Minquiers, I would have to say it is this ability to forget the nine to five, render your mobile useless, and live by the tide that draws me back, time and time again.

(Ostroumoff & Painter, pers. comm & In litt., 2020)

Martin Richardson

My family first became involved with Les Minquiers through my uncle Vivian [Richardson] as he acquired a hut on La Maîtresse Île in the 1930s, and my father [Advocate Dennis Richardson] subsequently bought a small hut in pre-war days. There are photographs of my uncle and aunt and grandfather sunbathing in deckchairs on the slipway in those early days. I don't really know how much they used the huts before the war but on one occasion my father and a party of friends were stormbound for many days. As my grandfather became concerned he arranged for an aircraft to fly over the Minquiers to see whether they wanted help. There was no need so they laid out sheets and arranged them in the word OK. As soon as the weather improved they were able to return to Jersey, and their story of being stormbound and marooned on the Minquiers was covered by an article in the JEP

During the war, when the Germans had an outpost on the reef they removed timbers from the roofs of the other huts to supply fuel for their fires and to keep themselves warm in the winter. That is why the majority of the huts fell into disrepair and became ruins. After the war my father acquired a hut on the Écréhous, and as the Écréhous was an easier place to sail to, he rather gave up on the Minquiers and eventually sold his hut [marked 'F' on the 1973 survey map, see Fig. 2)] to Peter Baker for £10. Peter never did anything with the hut but mentioned to one of his good friend's Advocate Bob Vibert, that if he, or any member of his family, wanted to rebuild the ruin they would be welcome to do so. Bob Vibert's daughter, Juliet, and her husband Richard Gillam, were already very enthusiastic about the Minquiers so were quick to take advantage of the offer and rebuilt the ruin. Many years later when Peter Baker died, it was a rather nice touch that he had left his hut to Juliet. So that meant the only hut the Richardson family had an interest in was a ruin owned by my Uncle Vivian Richardson.

Because my father had given up going to the Minquiers and spent time on the Écréhous instead, I had never been to the Minquiers until I returned to live in Jersey after ten years in the army when I was in my early thirties. One evening, while playing bridge with Uncle Vivian, who was rather short of money at the time, he asked me whether I would like to buy his hut. At that time I was an articled clerk with Coopers and Lybrand with no money at all, with three children, and I had to say no. But we were very friendly with Peter Newbold who was keen to buy a hut on either the Écréhous or the Minquiers. When I told my father about Uncle Vivian's offer, and that another person was interested, he decided to buy the ruined hut from his brother for, in hindsight, he had rather regretted having sold his hut on the Minquiers. It has subsequently proved to have been a wonderful acquisition to keep in the family.

At the end of the 1970s and early 80s Richard Falle and others were showing an interest in acquiring and restoring a property on La Maîtress Île. My father took the opportunity to combine resources with them and rebuild Uncle Vivian's hut. It was one of the last projects my father did. [marked 'J" on 1976 survey map]. Richard Falle's syndicate and my father were fortunate to have the use of the States' tug, the Duke of Normandy, to convey all the building materials and provisions out to La Maîtresse Île. My father employed Bob Hordle, a small-time builder from Grouville, to stay out there to rebuild the hut, and he soon became a complete convert to the Minquiers. The front of our hut faces west, whereas the back of the house looks onto the open space which is known as 'The Refuge'. There is a door to this that used to be unlocked, so if you were marooned on the Minquiers you could shelter there.

Our family still owns huts on the Écréhous. When my father bought his hut, the lady from whom he was buying was adamant that a particular person who already owned a number of huts was to be prevented from being able to buy any more properties. Therefore, as a part of the deal my father had to purchase a second plot next to the old Monastery on the reef. It was there for years as a ruin and we used to go over and inspect the ruin! There is a well on the island but it is pretty brackish and I don't think that anybody has used it for a century.

The Minquiers reefs are very wild and rugged and thanks to the quarrying that went on [during the end of the 18th and early 19th century], the rock formation is rather like a moonscape. It is also quite enchanting for once you get there you have a sense of achievement and a sense of isolation. It is not a place to go in bad weather, and it is always preferable to choose a sunny day. A sail there with a good picnic makes a marvellous day out. You're out in the elements all day, occasionally we have had to picnic in the hut but not very often. Unlike many who visit, I do not fish on the reef but have occasionally been ormering. On very low spring tides, by the States' mooring, I have collected clams. As the rising tide comes across the beach it is extraordinary to see the way the clams suddenly pop-up out of the sand, like magic. But I am not a keen fisherman, and when I go the reef it's more for a good lunch and to enjoy the scenery. When we first visited La Maîtresse Île we used

to clear the Tree Mallow from in front of the hut before realising that this wasn't a good thing to do. Now we are very keen to preserve the Mallow to ensure that it can continue to benefit the numerous migrant birds that pass through the Minquiers.

Nowadays I don't go to the Minquiers as much as I would like to. When the children were quite young, I used to spend at least one weekend there, usually in August and stay in the hut. I don't think we ever borrowed a hut apart from once using Juliet Gilliam's hut when the children all got very excited because they found a spider in it! For these weekend visits I chose a neap tide as this was easier for mooring my boat and amateur fishermen were less interested in staying which meant we had the reef to ourselves. It is really remarkable when you think that in the busy holiday season all around the Bay of St Michel, there are crowded beaches and harbours whilst the Minquiers remained untouched. But now I just go down on day trips. When the winds right, and the days right, it makes a very good three-hour sail. When we get there, we take the table and chairs out of the hut, carry them up onto the helicopter pad. We've had some wonderful lunches there with fine wines and watch the reef open-up to the west as the tide falls. It's a marvelous scene and I have never had any member of my crew complain for they have always thoroughly enjoyed the day.

Occasionally, there are visiting yachtsmen but only a few, and the reef hasn't really caught on like the Écréhous which is far easier to get to and unlike the Écréhous it is not in easy reach of marinas, such as Cartaret, and hence become crowded in the summer. Also, if you go to Chausey they have a great procession of tourist boats arriving every day, and the whole island becomes swamped with people - it is best to visit Chausey out of season. Fortunately, the Minquiers are not on a regular route and I think what is so marvellous is that there are only a few of us who use it and very few visitors. Therefore, we are most fortunate that La Maîtresse Île is situated in the Bay of Mont St Michel with its impressive big tidal range, which is slightly bigger than in Jersey, and due to the dangers of the surrounding reefs is more difficult to visit. The offshore reefs of Les Minquiers represent a spectacular environment and as hut owners we are most fortunate to have had the opportunity to witness, over many years, such a unique isolated moonscape of rocks and such a marvellous diversity of seascapes.'

In July 2006 the well-known Jersey artist, Nick Romeril, spent two weeks on the Minquiers, six days by himself and spending the majority of his time painting. After his return to Jersey he said: 'It is really a beautiful place and it changes all the time, although that can make painting a little difficult. But it was really, really, refreshing. It's funny how you get controlled by the elements. You have to take different things into consideration - the time of the day, the weather, the tide, whether the wind's coming in from the west. You feel very humble because you are not in control.' (after Lewis, 2006).

The landscape artist, Peter Collyer, well captures some of the delights of spending a few days on La Maîtresse: 'We lounge in soporific splendor. The sea rises, the sea falls. From their roof-top perches the great Black-backed Gulls taunt the Herring Gulls on the rocks below who squawk back... Cormorants and shags stand

in small groups as if waiting for a bus to come along. common terns, squabble over sand eels as a group of

Oystercatchers circle as if preparing to perform a 'Red Arrows' display. A mating pair of Clown-face Beetles shunt back and forth across a rock (Collyer, 2000).

(Richardson, pers. comm. 2010)

BIBLIOGRAPHY

Andreae, T. & Gee, J. (1998). 'Conquered Islands, Best of British. The Union Jack is Unfurled Over the Islands Again As We Reclaim the Minquiers'. Jersey: *Jersey Evening Post*

Audrain, T. (1997). 'Minquiers and Ecréhous Building to be Restricted', 5 Apr. Jersey: *Jersey Evening Post*

Ball, S. (2007). 'French Seals Seek Love in British Waters' [by Paris correspondent]. London: *The Times*

Binney, F. (2016). 'Les Minquiers'. 5 colour images, pp. 11-12, Autumn. Jersey: Société Jersiaise Newsletter

Blashford-Snell, J. (2020). 'Jersey Wrecks Explored', Jersey: *Rural Magazine*

Boland, H. (1904). 'Les Îles de la Manche'. Paris: Hachette

Brannan, J. (2003): 'Minquiers and Ecréhous'. 100, V11, 119 – 121, Oct. Jersey: *Channel Islands History Journal*

Butlin-Baker, H. (1981). 'Southern Outpost of Jersey: The Minquiers, an Offshore Island'. 5 b. & w. images, pp. 12-15, Oct. Jersey: *The Islander*

Carnegie, P. (2006). 'The Channel Islands – Les Ecréhous and Plateau des Minquiers', pp. 131-140. Published for Royal Cruising Club Pilotage Foundation (fully revised 2nd edition). St Ives: Imray Laurie Norie & Watson Ltd

Chambers, P. (2008). *Channel Island Marine Molluscs – An illlustrated Guide to the Seashells of Jersey, Guernsey, Alderney, Sark and Herm*. Jersey: Charonia Media

Chambers, P. Binney, F, and Jefffrey's G. (2016). *Les Minquiers: A Natural History*. Jersey: Charconia Media

Chambers, P. (2017). 'New Focus – Environmental Gems that are Home to 750 Species'. 2 colour images, p. 8-9, 30 Sep. Jersey: *Jersey Evening Post*

Chiang, T. (2011). French Study into Minquiers Dolphins. 1 colour image, 29 Sep. Jersey: *Jersey Evening Post*

Chiang T. (2013). News - 'Deputy Wants a Ban on Offshore Reef Wind Farms'. 1 colour image, p.8, 15 Feb. Jersey: *Jersey Evening Post*

Chiang, T. (2013). 'Planning Happy with Offshore Weather Masts'. But Scheme Angers Some Environmental Groups. 1 b. & w. image, 23 Apr. Jersey: *Jersey Evening Post*

Chiang, T. (2013). Offshore Reef Weather Station Up and Running. P. 3 24 Sep Jersey: *Jersey Evening Post*

Clarke, J. (2006). 'Archaeology – Brief Visit to the Minquiers Where Two Late 18th Century fisherman's Huts are Being Rebuilt' [Guiton & Family].1 colour image, 45, 3. Jersey: Société Jersiaise Newsletter

Clarke, J. (2009). 'Excavation at Les Minquiers 2006'. Vol. 30 (1), 103-112. Jersey: Société Jersiaise, 134th *Annual Bulletin*

Collyer, P. (2000). Plateau des Minquiers. In *South by Southwest Painting the Channel Islands*. Pp. 149-154. Bradford-on-Avon: Thomas Reid Publication

Cooke, P. (1967). The Last Hours of the Flying Boat 'Iona'. Vol. 3 (28), 42-43, Jul. Jersey: *Jersey Topic Magazine*

Coon, D. (2001). Maritime Connections: Living from the Reef. In *Grouville, Jersey – The History of a Country Parish*, ed, R. Anthony,pp. 100-105, 5. Jersey: Parish of Grouville

Courault, P (2000). The Channel Islands; Les Minquiers. 2 colour images (by Didier Decoin) Les Minquiers. Cherbourg: EDF, GE & Jersey Electricity Company

Crosby, A. (1999a.). Minquiers 'Safe Haven' for Grouville?' 1 colour Image (Peter Mourant) p. 3, 19 May Jersey: *Jersey Evening Post*

Crosby, A. (1999b.). 'A Corner that is Forever Grouville - The Minquiers Maybe the Subject of Renovation Work – by the Army'. 2 b. & w. images (Peter Mourant), pp. 32-33. Jersey: *Jersey Evening Post*

Cudlipp, R. (2016). 'New Focus – Where Conservation is a Perfectly Natural Fit, Les Minquiers', p. 9, 30 Jun. Jersey: *Jersey Evening Post*

Culley, M, Farnham, W. Fletcher, R, and Thorp, C. (1993). 'The Marine Ecology of Maîtresse Ile, Les Minquiers', Pp. 1- 49, July 1992. Unpublished Report, Hayling Island: The University of Portsmouth

Cumberlidge, C. (1990). 'Graveyard Cruising' [Les Minquiers], 2 b. & w. Images. pp. 55-56, Mar. London: *Motor Boat and Yachting*

Cummings, J. (1999). 'Minquiers Business - This is the Home that Frank Lawrence Built on the C. I. Reef', pp. 16-17, 15 May. Jersey: *Jersey Evening Post*

Daily Mirror (1929). Comedy of "French Invasion" of a Little Island. Twelve Men and a Dog Defend Union Jack, 4 b. & w. images, pp. 12-13, July. London: *Daily Mirror*

Daily Mirror (1929). French "Invasion" of Minquiers Island, 6 b. & w. images.17 Jul. London: Daily *Mirror*

Daly, S. (2004). *Wildlife of the Channel Islands*, (Foreword by Lee Durrell). Wiltshire: Seaflower Books

Davies, W. (1971). *Fort Regent: A History*, p. 65. Jersey: Privately Publ.

De Gruchy, G. (1938). "Les Minquiers'. Vol. XII (3), 297-298. Jersey: Société Jersiaise *Annual Bulletin*

De Saint-Père, J. (1939). 'A Qui Sont Les Minquiers?' Rennes: L'Quest-Dimanche

Dobbie, P. (1984). 'Another Victory for the Flag ... Island Raid by Lunch Party', p. 3, 3 Jun. London: *Mail on Sunday*

Dobson. R. (1952). *The Birds of the Channel Islands*. London: Staples Press

Dole, H. (1997). *Victor Hugo's 'Ninety-Three*. Translated by Helen B. Dole. [Refs. to Minquiers pp.20 & 63]. Milton Keynes: Rough Draft Printing/Watch Making Publishing

Eaves, K. (2015). 'Visitors Urged Not to Cause Damage at Reefs'. 1 b. & w. image, 24 Apr. Jersey: *Jersey Evening Post*

Edbrooke, D. (2016). 'Everything You Ever Wanted to Know About the Minquiers'. 1 colour image, p. 17, 2 Nov. Jersey: *Jersey Evening Post*

Falle, P. (1998). 'Jolly Good Show!' 5 Oct. (Picture: Award Presentation). Jersey: *Jersey Evening Post*

Falle, P. (2005). 'Last Item on the Agenda'. 1 b. & w. image (members of the Harbours and Airport Committee on a visit to the Minquiers in 1958), pp. 8-9, 5 Dec. Jersey: *Jersey Evening Post*

Falle, R. (2001). 'Maritime Connections: Les Minquiers'. In *Grouville, Jersey - The History of a Country Parish*,ed, R. Anthony, 5: 91-99. Jersey: Parish of Grouville

Fauchon, M. (1975). 'La question des Minquiers et des Ecréhou'. Revue de l'Acranchin et du Pays de Grouville. Tome 52 No. 282

Finlaison, M. (1980). Archeological Section Report. , Vol. 105, 372-373. Jersey:Société Jersiaise *Annual Bulletin*

Fischer, P. & E. (1926). 'Molluscques Recoltes aux Minquiers'. Vol. 70. 57-61. France: Journal de Conchyliologie

Fremine Ch. et A. (1888). 'Les Normands dans Les Îles de la Manche', Paris: Picard et Kaan

Godfray, A. (1929). Archaeological Researches at the Minguiers. Vol. 11 (2), 193 -199. Jersey: Société Jersiaise *Annual Bulletin*

Godfray, A. (1931). Archeological Report, Les Minquiers. Vol. 11, 308-309. Jersey: Société Jersiaise *Annual Bulletin*

Graham, C. (1954). Ornithological Section Report for 1953. 2 b. & w. images, Vol. 16 (11), 128-137. Jersey: Société Jersiaise *Annual Bulletin*

Grigson, G. (1954). 'A Reef of Islets, Minquiers Reef'. In *People, Places, Things and Ideas*. Pp. 214 & 231. London: Grosvenor

Grove, J. (2017). 'Community - Landmark Ruling Over Reefs'. 2 colour, Images, p.20, 19 Jun. Jersey: *Jersey Evening Post*

Gruchy, G. de (1938). Les Minquiers: Exttraits des Rôles de la Cour et du Fief et Siegneurerie de Noirmont. Vol. VIII (3), 297-298. Jersey: Société Jersiaise *Annual Bulletin*

Guéguen, N. (2017). 'Les Minquiers: Plus grands que les villes britanniques! Section Bilologie Marine de Societé Jersiaisee'. 1 b. & w. image, 20 Apr. Jersey: *Jersey Evening Post*

Heath, R. (2019). Historic Headlines - The Seaplane That Was Lost at Sea With All On Board, 4 b. & w. images, pp. 12-13, 13 Jul. Jersey: *Jersey Evening Post*

Heath, R. (2020). Night Rescue As Yacht Sinks Off Reef. 1 colour image, p. 6, 2 Jun. Jersey: *Jersey Evening Post*

Heuston, A. (2003). 'Ramsar Status for Reefs? Convention: Ecréhous and the Minquiers Could be Considered for Future International Recognition', p. 7, 22 Nov. Jersey: *Jersey Evening Post*

Hill, L. (1986). 'Geological Survey of Le Plateau des Minquiers'. Vol. 22 (4), 384. Jersey: Société Jersiaise *Annual Bulletin*

HM Stationery Office (1952). Agreement Regarding Rights of Fishing in Areas of the Ecréhos and Minquiers, London, 30 Jan. 1951. Treaty Series No. 4, Caid. 8444, 1-7, 2 maps (3367). London: HM Stationery Office

Hunt, P. (2012). 'Temps Passé: Jersey's Southern Outpost, Jeremy Mallinson has recently written a book about the Minquiers'. 2 b. & w. images, p, 23, 15 Mar, Jersey: *Jersey Evening Post*

Hutchison, J. (2011). 'Protection Plan for the South-East'. 1 colour image, p. 5, 4 Feb. Jersey: *Jersey Evening Post*

Hutchinon, J. (2012). 'Protest Over 30-ft Mast for Ecréhous'. 2 colour images, pp. 1-2, 23 Jun. Jersey: *Jersey Evening Post*

I.C.J. (1953). 'The Minquiers and Ecrehos Case (France/United Kingdom) Judgement of November 17th, 1953'. Vols. 1 & II. Netherlands: I.C.J.

Innes, H. (1956). *The Wreck of the Mary Deare*. London: Collins

James, J. (2007). 'La Maîtrese Sauvage', 06/07: 32-34. Jersey: Indulgent Traveller

Jamieson, A. (ed). (1986). *A People of the Sea: The Maritime History of the Channel Islands*, p. 270. London: Methuen

Jefferson, D. (1985). 'Britain's Southern Shoals: David Jefferson Visits Plateau des Minquiers in the Channel Islands' 1 b. & w. image & map, Vol. LVII, 485-495, Sep. London: *Geographical Magazine*

JEP (1936). 'Disappearance of Sea Plane *Cloud of Iona*'. 1 b. & w. image, p. 4, 1 Aug. Typescript. Jersey: *Jersey Evening Post*

JEP (1937). 'Cloud of Iona' The Ministry's Inquiry (throws little light on the disaster), 15 Feb. Jersey: *Jersey Evening Post*

JEP (1951). 'Good Publicity for Jersey - Ormering Expedition to Minquiers as Week-end Highlight'. 2 b. & w. images, p. 10, 24 Apr. Jersey: *Jersey Evening Post*

JEP (1953). Bristol Trader Strikes Minquiers - Lifeboat Tender Goes Out to Assistance. 5 b. & w. images, 15 Dec. Jersey: *Jersey Evening Post*

JEP (1954). 'The Duchess Returns'. 1 b. & w. image. Jersey:*Jersey Evening Post*

JEP (1956). 'French Fishermen Marooned on Maitresse Ile – Rescuedby Cancale Yacht'. 1 b. & w. image, 18 Aug. Jersey: *Jersey Evening Post.*

JEP (1969). How Much is a Cannon Worth? 1 b. & w. image p.6, 16 Jan. Jersey: *Jersey Evening Post*

JEP (1976). 'Minquiers', 2 b. & w. images, 18 Jun. Jersey: *Jersey Evening Post*

JEP (1999). 'The State: Planning's Minquiers Visit "Was Part of Officers'Professional Duties"' I b. & w. image, p. 29, 10 Jun Jersey: *Jersey Evening Post*

JEP (2004). 'Temps passé - Study Trip: Members of the Société Jersiaise Bird Section Erect a Net Trap on the Minquiers in 1952'. 3 Mar. Jersey: *Jersey Evening Post*

JEP (2008). 'What's On: The Week: Talking Heads - Ref. to Archaeology lecture on the pre-history of Les Minquiers and later human activities' by John Clarke, p. 21, 12 Jan. (Picture: Société Jersiaise circa. 1928 of the Maîtresse Île). Jersey: *Jersey Evening Post*

JEP (2010). A Weather Eye on Jersey - Ref. to the Sandbank Close to the Fourchi Rouge rock, Les Minquiers.1 colour image by Paul Chambers, p. 39. 9 Sep. Jersey: *Jersey Evening Post*

JEP (2013). News – Jewels of the Sea.1 colour image [huge sandbar], p.8, 20 Jun. Jersey: *Jersey Evening Post*

JEP (2013). 'Love on the Rocks! – two local suitors decided that Jersey's Offshore reefs made for the perfect location'. 1 colour image, p. 3, 10 Jul. Jersey: *Jersey Evening Post*

JEP (2013). 'Weather & Tides – A Weather Eye on Jersey. Sunrise Seen From the Minquiers'. 1 colour image, p. 20, 18 Sep. Jersey: *Jersey Evening Post*

JEP (2013). 'Temps passé – Minquiers Adventure'. 1 b. & w. image, p. 14, 23 Sep. Jersey: *Jersey Evening Post*

JEP (2013). 'Weather & Tides – A Weather Eye on Jersey'. 1 colour image, p. 20, 2 Oct. Jersey: *Jersey Evening Post*

JEP (2015). 'Eight Rescued Off Reef After Dinghy "Loses Air"'. 1 b. & w. image, 18 Jul. Jersey: *Jersey Evening Post*

JEP (2015). 'Jersey through the Decades'. 1 b. &. w. image, p. 40, 22 Aug. Jersey: *Jersey Evening Post*

JEP (2015). 'A Helicopter Lands on the Minquiers in May 1999'. 1 b. & w. image, p. 37, 19 Sep. Jersey: *Jersey Evening Post*

JEP (2015). 'Reigning in the South – The flag is raised on Maitresse Ile'. 1 b. & w. image, p. 11, 22 Sep. Jersey: *Jersey Evening Post*

JEP (2016). 'Weather and Tides – A Weather Eye in Jersey'. I colour image of Les Minquiers , p. 25, 14 Jun. Jersey: *Jersey Evening Post*

JEP (2015). 'Homelife – Bought and Sold. Site visit to the Minquiers for the sale of l'Hospital, one of the small huts on the reef'. 2 colour images (Viberts). Jersey: *Jersey Evening Post*

JEP (2016). Pictures '80th Anniversary of a Tragedy – A Flight Lost at Sea'. 7 b. & w. images, pp. 18-19, 28 Jul. Jersey: *Jersey Evening Post*

JEP (2017). News 'Reefs May Get More Protection', p. 7, 18 May. Jersey: *Jersey Evening Post*.

JEP (2018). 'Homelife, Bought and sold – A Small House on the Minquiers Reef Was Sold for £35,000'. I colour image, p. 50, Nov. Jersey: *Jersey Evening Post*

JEP (2019): Opinion- From the JEP's Photographic Archives. b. & w. image of important work being carried out by the bird section of the Société Jersiaise on the Minquiers (early spring bird ringing), p. 15, 28 Mar. Jersey: *Jersey Evening Post*

JEP (2019). B & W image of the Union flag being hoisted on On the Minquiers of Jerseyman Private R. Marquis in September 1945 p. 15, 9 May. Jersey: *Jersey Evening Post*

JEP (2019). 'Fishing Restrictions Approved', 28 Sep. Jersey: *Jersey Evening Post*

JEP (2019). 'A map showing the area covered by the Jersey National Park'. 3 colour images, p. 7, 27 Dec, Jersey: *Jersey Evening Post*

Jersey Post (2003). Jersey Offshore Reefs – First Day Cover - 5 stamps illustrating various flora found on the reefs, 2 depict cottages on Les Minquiers and flora. Jersey: *Jersey Post*

Jersey Times (1910). 'Missing Fishermen, Tug to the Rescue', 8 April. Jersey: *Jersey Times*

Jersiaise (1692). Archival Material – [Translated from French - References only]:

(1692). Dispute Between Dam de Samarès and the King. Re. Possession of wreckage on the Minquiers'. (Ref. S.B. 9: 18 a. & b. Shelf F7). Jersey: Société Jersiaise. (1692) Summons to Attorney General and Receiver to appear before Privy Counci.l Re. Deborah Dumeresque's appeal in her claim to wreckage on Les Minquiers. (Ref. S.B. 9: 18 a. & b. Shelf, F7). Jersey: Société Jersiaise

J.W.P. (1945). 'French Flag Hoisted on Minquiers – Incident of 1929' Recalle, p. 1, 11 Aug. Jersey: *Jersey Weekly Post*.

Jeune, P. & Macintyre, B. (1998). British Bobby Retakes Channel Isle, Bobby Retakes Reef for Queen, 1-2, 2 Sep. London: *The Times*

Jouault, N. (2001). 'Maritime Connections: The Sinking of the *Polka* and the *Superb*'. In *Grouville, Jersey - The History of a Country Parish*, ed, R. Anthony, 5: 106-109. Jersey: Parish of Grouville.

Jouault, N. (2010). Marine biology. P. 13 Autumn. Jersey: Société Jersiaise Newsletter,Vol. 53.

Kempster, H. (1981). 'Looking Through a Telescope'. A Story About the La Rocque Fishermen. 2 b. & w. images, pp. 1-4. Jersey: An unpublished manuscript

La Morandiere Ch. de (1956). 'Les Archipels Normands des Minquiers et des Ecréhou'. Etudes Normandes No. 75

Lasaygues, F. (2006). 'Anglo-Normandes Minquiers / Écréhou Voyage dans Les Îles de l'oubli'. 19 colour images by Patrick Courault, 12: 20- 35. RCS Caen: Au fil de la Normandie

Lawrence, F. (2010). *Minquiers Pilotage Notes*. 6 colour images, p. 73, (PP Produced by St. Helier Yacht Club and Port of Jersey. Jersey: P. J. News and Publisher

Le Gros, C. (1943). Traité du Droit Coutumier de L'Ile de Jersey. Jersey: Les Chroniques de Jersey, Ltd

Le Maistre, D. (2011). *Low Water Fishing: An Islander's Pursuit*. Bradford on Avon: Seaflower Books/Ex Libris Press

Le Moine Gray, A. (2009). *Jersey – Portrait of an Island*. 2 colour paintings of Les Minquiers, pp. 106-107. Bretagne: Cloître Imprimeurs, Saint-Thonan

Le Scelleur, K. (2000). The Evacuation of St. Malo. Published on behalf of the St. Helier Yacht Club. Jersey: P.J. News & Publishing

Le Sueur, F. (1976). *A Natural History of Jersey*. London: Phillimore

Le Sueur, F, Long, M, Long, R, and Paton, R. (1973). 'Report on a Visit to Les Minquiers by Joint Nature Conservation Advisory Body, to H.M. Receiver General'. 12 b. & w. images. Jersey: Société Jersiaise

Lebel-Jehenne, L. (1951). 'Demande de Concession des Minquiers' (1784). Article extrait de C 1206 Archives du Calvados. Grauville: La Pays de

Lemprière, R. (1970). *The Channel Islands*, p. 31. London: Robert Hale

Lemprière-Robin, R. & Falle, R. (1986). 'Fast Currents and Fistfuls of Rocks, Les Ecréhous and Le Plateau des Minquiers'. 7 b. & w. images, pp. 438- 440, 7 Aug, 8321. London: *Country Life*

Lewis, A. (2006). 'Painting Without Numbers'. Pp. 10-11, 25 Jul. Jersey: *Jersey Evening Post*.

Lowe, J. (2019). 'Invaders Claim British Rock for a Rogue French' "King. Pp. 10-11, 27 Oct. London: *Sunday Telegraph*

Macintyre, B. (1998). 'Channel Islet Seized by King of Patagonia' [by Paris correspondent]. London: *The Times*

Maguire, J, (2020). 'Catching of Bluefin Tuna Banned'. P. 5, 8 Apr. Jersey: *Jersey Evening Post*.

Masterman, P. (2017). Pictures – The Minquiers. 6 colour images, pp. 16-17 Jersey: *Jersey Evening Post*

Mauger, R. (1978). 'A Voyage to the Minquiers'. Pp. 10-11, 3 Mar. Jersey: *Jersey Evening Post*.

Mollet, J. (1894-97). French Claims to Minquiers. Newspaper Cuttings-Archives, 69-71 & 81. Jersey: Société Jersiaise

Morel, J, (2020). Feature: A Little 90s Dancefloor Magic Could Well Boost This Health Campaign. 4 colour images, p. 20, 25 Mar. Jersey: *Jersey Evening Post*

Morning News (1910). 'Disappearance of Boat (Les Minquiers), No Food or Water for Two Days'. Jersey: *Morning News*

Mourant, Arthur (1933). 'A Geology of the Ecréhous, Paternosters and Minquiers'. Vol. 12 (2),180-183. Jersey: Société Jersiaise *Bulletin*

Mourant, Arthur (1973). 'One Hundred Years of Jersey Geology and Geologists' (Minquiers p. 55). Jersey: Société Jersiaise

Mourant, Andrew (1992). 'First Flushing Toilet Panned by Islanders' [Les Ecréhous], 30 Aug. London: *Sunday Telegraph*

Mourant, P. (1999). 'Taking to the Air for Some Minquiers Business'. 1 b. & w. image of Les Minquiers at high tide, 15 Jul. Jersey: *Jersey Evening Post*

Mourant, P. (2007). 'Reef Encounter'. 7 b. & w. images, pp. 24-25, 11 Sep. 24 Jersey: *Jersey Evening Post*

Neptune (1989). 'Les Îles du Littoral Français Les Minquiers, un Sacré Grain de Sable', *Revue Nautique*, P. 219. Granville: Edition Glénat

Parish of St Brelade (2009). A Report from the Action Group on Ramsar

Status for the West End of St. Aubin's Bay – as amended. (Report tabled and accepted by a St. Brelade Parish Hall Assembly with the intention for the report to be forwarded to the appropriate States authority by the Connétable, in order for its recommendations to be incldeed in the Island Plan Policy document. Jersey: Parish of St. Brelade

Petters, L. (1998a). Protest: Reef is Claimed … Again! Minquiers 'Farce' Gets an Encore, Sep. Jersey: *Jersey Evening Post*

Petters, L. (1998b). Silly Season Brings on Touch of Old Minquier Business. 4 b. & w. images by Christian Keenan, 2 Sep. Jersey: *Jersey Evening Post*

Potigny, F. (2019). 'Minquiers Invaders Speak Out: "Brexit Made Us Do It"'. 4 colour images, 29 Oct. Jersey. *Bailiwick Express*

Quérée, B. (2004). Conservation: Island Celebrates the 32nd Anniversary of Ramsar Convention – Walks and Talks Planned to Mark Wetlands Week, p. 11, 30 Jan. [Minquiers referred to in text]. Jersey: *Jersey Evening Post*

Quérée, B. (2012). 'On a Fine Day, the Minquiers Look Like a Scene From Pirates of the Caribbean B. Q. discovers'. 1 b. & w. image, 10 Nov. Jersey: *Jersey Evening Post*

Ramsar (2009). The Annotated Ramsar List. Switzerland: Ramsar Convention Secretariat, Gland

Ramsar Convention (2010). Comments on the Ramsar Consultation Document to Have a Relevance to the Current Status and Future Management of Jersey's Ramsar sites, 4 pp. Jersey: Marine Biology Section, Société Jersiaise

Reed, B. (1989). 'Minquiers, Jersey's Southern Outpost'. Vol. 3 (5),191-196, Sep. Jersey: Jersey Society in London *Bulletin*

Richardson, D. (1952). 1929 - The French Claim to Les Minquiers (Maitre Isle) 1951-1964. Jersey Society in London *Bulletin*

Robinson, A. (1959). Geological Report. Incl. Geological Map of the Minquiers, copied from a paper by Messrs Grainda and Roblot in the Bulletin of the French Geological Society, Vol. 17 (3), 217-218, 1957. Jersey: Société Jersiaise Annual *Bulletin*.

Robson, M. (1979). *Guide Nautique des Îles Anglo-Normandes*. Paris: Arhaud

Roche, A. (1953). *The Minquiers and Ecréhou Case*. Paris: Librairie Droz, Genève et Minard.

Rodwell, R. (1996). *Les Écréhous, Jersey - The History and Archaeology of a Channel Island Archaepelago*. Jersey: Société Jersiaise

Rubinstein, B. (1976). 'Des Minquiers Nommé Désir'. 13 b. & w. images by author. Paris: *Yachting Magazine*

Rutherford, W. (1976). Jersey, p. 19. Newton Abbot: David & Charles.

Rybot, N. (1928). Plateau Des Minquiers, Mâitresse Île, Jersey. A 1928 Sketch of Island and a List of Owners of the Various Huts [15]. Vol. 17 (4).15, 26 & 27 Jul 1973. Reproduced by the Nature Conservation Advisory Body. Survey Jersey: Société Jersiaise Annual *Bulletin*

Sabon, P. (1929). 'L'Angleterre et la France Revendiquent l Propriéré d'un Îlot des Minquiers', 12 Juillet. Paris: Liberté

Sibey, A. (2013). 'Jersey Should Explore Idea of Offshore Wind Farms.' But Environment Minister Has No Plans to Site Farms on the Ecréhous or Minquiers, 5 Apr. Jersey: *Jersey Evening Post*

Stephenson, L. (2013). News – Plans for masts on Minquiers and Ecréhous are scrapped. 3 color images, p. 5, 25 Jan. Jersey: *Jersey Evening Post*

Stephenson, L. (2013). News – 'Plans to Site Weather Masts Away From Reefs Welcomed'. 2 colour images, p. 5, 1 Feb. Jersey: *Jersey Evening Post*

Stephenson, L. (2016). 'Island Service That's at Your Convenience'. 1 colour image, p.12, 4 Jan. Jersey: *Jersey Evening Post*

Shipley, R. (1996). 'On the Rocks Without Ice?' 4 colour images, pp.14-15, 21 Sep. Jersey: *Jersey Evening Post*

Shipley, R. (2011). 'What Lies Ahead for Our Green and Pleasant Land?' pp. 10-11, 6 Apr. Jersey: *Jersey Evening Post*

Shipley, R. (2011). 'In Praise of a Rocky Outpost'. 2 colour images, 13 Jul. Jersey: *Jersey Evening Post*

Simon, D. (2010a). 'New Group Keeps a Close Eye on Care of Eco-sites'. 1 colour image, p. 5, 8 Mar. Jersey: *Jersey Evening Post*

Simon, D. (2010b). 'Ramsar Authority Begins Protection Debate'. 2 b. & w. images, 4 May. Jersey: *Jersey Evening Post*

Sinsoilliez, R. (1995). *Histoire des Minquiers et des Écréhou*. St. Malo: Bertrand de Quénetain

Smith, D. (1978). 'Looks Back to the Events of 25 Years Ago - The Fight for Sovereignty of the Minquiers', 19 Sep. Jersey: *Jersey Evening Post*

Spruyt, J. & Baker, H. (1997). 'Wide World: Island Life, Britain's True Land's End', 1 b. & w. image, pp. 46-47, Mar. London: *Geographical Magazine*

States of Jersey (1947). Registre de Pilotes (States Harbours Department). Jersey: States of Jersey

States of Jersey (1994). Fishing Regulations: Fishing for Ormers. Jersey: Sea Fisheries (Jersey) Law. Jersey: States of Jersey

States of Jersey (2008). Making the Most of Jersey's Coast: Integrated Coastal Zone Management Strategy. Pp. 1-39, March. Department of Planning and Environment. Jersey: States of Jersey

States of Jersey (2011). Jersey's South East Coast Ramsar Management Plan. Pp. 1-38, 1 Feb. Jersey: Department of Planning and Environment. Jersey: States of Jersey

Stentiford, M. (2004). 'Nature: Ramsar Designation, Safer Havens', p. 33, 20 Mar. Jersey: *Jersey Evening Post*

Stevens, J. & Jee, N. (1987). *The Channel Islands: The Bailiwick of Jersey* Les Minquiers, p. 54 (Introduction by John Arlott). London: New Shell Guides

Syvret, A. (1995a.). Marine Biology. Vol. 29, 40-42, 130th *Annual Bulletin*, Jersey: Société Jersiaise

Syvret, A. (1995b.). Marine Biology. 1 colour image Vol. 42, 6. Jersey: Société Jersiaise Newsletter

Syvret, M. & Stevens, J. (1981 & 1998). *Balleine's History of Jersey* - The Coming of Steam, p. 471, (Revised and enlarged edition of G. Balleine's 1950 first publication). Chichester: Phillimore & Co. Ltd

Thelwell, P. (2019). News Focus: 'Room to breathe - The States Have Agreed to Invest £100K in the Jersey National Park'. 2 colour images, p. 6. 27 Dec. Jersey: *Jersey Evening Post*.

The Isle (2010). Saltwater in his Blood. Captain Frank Lawrence is a Senior Seaman of a Seafaring Family, Vol. 2 (15), 9, Mar. Jersey: People, The Isle (Jersey's monthly independent information source

Thomas, R. (1992). *Lest We Forget, Escape and Attempted Escapes from Jersey During the German Occupation*. Jersey: La Haule Books

Thompson Estates (2019). 'Ever Visited the Minquiers? Well, Now Is Your Chance!' 1 colour image, front cover & p. 3, Sep. Issue 180. Jersey: Property Monthly

Trevor, W. (1928). Map 5. 'The Channel Isles, Showing the Ten Fathom Line'. Vol. 11, 84 - 85. Jersey: Société Jersiaise *Annual Bulletin*

Tweedie, N. & Jeune, P. (1998). 'Lone Policeman Raises the Flag to Repel In Invaders', p. 10, 2 Sep. London: *Daily Telegraph*

UK Hydrographic Service (1999-2010). Wreck Reports (8), H.M. Stationary Office, http://www.recksite.eu/ukhoDetails.aspx.

Young, G. (2016). 'Read All About the Minquiers Magic'. 4 colour images, p.16, 10 Nov. Jersey: *Jersey Evening Post*.

About the author

Jeremy Mallinson OBE, DSC (HON), FRSB, FRGS

In March 1951 he came to live in Jersey with his parents and elder brother, and while he was at The King's School, Canterbury. In May 1959, after over two-year service in the Rhodesia & Nyasaland Staff Corps he joined the staff of Gerald Durrell's newly formed Jersey Zoo. He subsequently held the posts of Deputy and Zoological Director of the Jersey Wildlife Preservation Trust. And after the death of his mentor, Gerald Durrell, in 1995 he was appointed Director of the renamed Durrell Wildlife Conservation Trust.

Since his retirement in 2002, he has continued to support and to be involved in a number of international conservation projects in Brazil. He is an ardent advocate for the prevention of any further degradation of Jersey's coastal ecosystems (including its offshore reefs of Les Minquiers and Les Écréhous), and the need to promote further the conservation of the islands' fauna and flora.

He remains a staunch supporter of the international conservation role of the Durrell Wildlife Conservation Trust, and is a keen follower of the National Trust for Jersey's Coastal Campaign. In 2007 he became chairman of the St Aubin's Anti-reclamation Action Group.

More books by Jeremy Mallinson:
Okavango Adventure
Earning Your Living with Animals
Modern Classic Animal Stories – chosen by
The Shadow of Extinction
The Facts About a Zoo
Travels in Search of Endangered Species
'Durrelliana' – An Illustrated Checklist
The Count's Cats - a novel
The Touch of Durrell: A Passion for Animals
Someone Wishes to Speak to You - a novel
On the Subject of Relationships - a novel

Complete list of Seaflower Books 2020:

CHANNEL FISH by Marguerite Paul	£11.95
CHEERS! Drinks & drinking in Jersey through the ages by Alasdair Crosby	£9.95
EXOTIC GARDEN PLANTS IN THE CHANNEL ISLANDS by Janine Le Pivert	£9.95
ISLAND DESTINY by Richard Le Tissier	£6.95
ISLAND KITCHEN by Marguerite Paul	£11.95
JERSEY JAUNTS by John Le Dain	£5.95
THE JERSEY LILY by Sonia Hillsdon	£5.95
JERSEY OCCUPATION DIARY by Nan Le Ruez	£9.95
JERSEY RAMBLES by John Le Dain	£6.95
JERSEY: SECRETS OF THE SEA by Paul Darroch	£11.95
JERSEY: THE HIDDEN HISTORIES by Paul Darroch	£11.95
JERSEY WAR WALKS by Ian Ronayne	£8.95
JERSEY WITCHES, GHOSTS & TRADITIONS by by Sonia Hillsdon	£6.95
JOURNEY ROUND ST HELIER by Robin Pittman	£7.95
LIFE ON SARK by Jennifer Cochrane	£5.95
MINED WHERE YOU WALK by Richard Le Tissier	£6.95
THE OTHER JERSEY BOYS by David Knight	£9.95
PHILIP DE CARTERET R.N. by Jane Ashelford	£9.95
PROMISES NOT FORGOTTEN by Gerald Breen	£11.95
THE POOR SHALL INHERIT Daff Noel	£6.95
WILDLIFE OF THE CHANNEL ISLANDS by Sue Daly	£14.95
WISH YOU WERE HERE by John Le Dain	£7.95

Please visit our website for more details: **www.ex-librisbooks.co.uk**
SEAFLOWER BOOKS may be ordered through our website using Paypal
We send books post-free within the UK and Channel Islands
SEAFLOWER BOOKS are also available via your local bookshop or from Amazon.com

SEAFLOWER BOOKS
www.ex-librisbooks.co.uk